# Stability and Change in High-Tech Enterprises

At the turn of the millennium the nature of work is undergoing major developments during one of the most rapid periods of economic and technological change experienced in the modern world.

This book examines how firms, as particular economic institutions, live through and experience change. Four high-technology firms are studied. The author provides a rich analysis of the routines of chosen employees and shows how the people are continually engaged with change. The explanations which develop provide new foundations for the understanding of firms in any sector. They also emphasise the importance of empirical evidence, rule-based action and the need to recognise the historical, social and interpretative contexts of the working environment.

The book's primary contribution is the development of a broader concept of routine, in particular the identification of routine practices at a strategic level and the demonstration that such practices can incorporate change. The analysis identifies the significance of technology in economic change and confirms the value of an institutional approach to the understanding of a firm's behaviour.

**Neil Costello** is Head of Economics at the Open University. He has held a number of senior posts at the OU and elsewhere and has undertaken consultancy in international development. His teaching and research have increasingly been based around developing an understanding of real-life cases using explanations drawn from a wide range of theoretical positions.

## Routledge Studies in Business Organizations and Networks

# Stability and Change in High-Tech Enterprises

## Organisational practices and routines

Neil Costello

London and New York

First published 2000
by Routledge
11 New Fetter Lane, London EC4P 4EE

Simultaneously published in the USA and Canada
by Routledge
29 West 35th Street, New York, NY 10001

*Routledge is an imprint of the Taylor & Francis Group*

Typeset in Sabon by Keystroke, Jacaranda Lodge, Wolverhampton
Printed and bound in Great Britain by MPG Books Ltd, Bodmin

*British Library Cataloguing in Publication Data*
A catalogue record for this book is available from the British Library

*Library of Congress Cataloging in Publication Data*
Costello, Neil, 1947–
Stability and change in high-tech enterprises : organisational practices and
routines / Neil Costello.
p. cm.
Includes bibliographical references and index.
ISBN 0-415-23121-3
1. High technology industries—Management—Case studies.
2. Organisational change—Case studies.  3. Small business—
Management—Case studies.  I. Title.

HD62.37 .C67 2000
620'.0068—dc21    00–038249

ISBN 0-415-23121-3

# Contents

# Preface and acknowledgements

The ideas presented in this book have their origins in the distant past. I have been interested in taken-for-granted behaviour in organisations since the late 1960s when I worked for a major motor manufacturer which did not seem to behave according to standard maximising principles. Subsequently, I was a member of a research team looking in detail at investment decisions in the fuel and power industries, in which one of my tasks was to try to map the concerns of engineers onto the concepts of economists. The value of real-life case studies in teaching and research developed slowly from that: case studies in which theory was important but needed to prove its value by providing a cogent analysis of each specific case, rather than building the case to illustrate its own apparent power. My teaching, consultancy and research have become focused around specific cases since then. The cases provide building blocks for theorising, however, and insights into organisational change, and are not simply detailed descriptions of specific events and institutions.

In the early 1990s an opportunity arose to work on a project which looked at change in organisations; it was set in a broader context of discourse in economics and examined the claims being made for the importance of the new information technologies in social and economic change. The project enabled me to carry out longitudinal analyses of firms for the first time, and to look at change over time rather than as a snapshot at a particular moment. Much of the evidence presented here rests on interpretations of conversations and interviews with people working in the firms. In the chapters which use the case material in detail (Chapters 4–7) the words of the participants are used frequently to give them a part in telling the story. The quotations should be read as part of the text. They are used to develop the argument and reflect the research method adopted here. Part of that is to acquire a sense of change as it appears to those involved with it and to use, as much as possible, their agenda and their perceptions.

The intention is to look at the way in which sense making takes place in firms such as those studied here, to look at the way in which people live with change. Each of the firms was involved in one way or another with high technology. In some respects they lived (rather than lived *with*) change and also lived high tech. The individuals in these companies were engaged with and enduring change. It was a central part of their working lives. I hope the case studies are able to deliver a sense of what that felt like for the participants. It is an important part of our experience at the turn of the millennium and to that extent the statements and reflections of the participants in the research can stand as statements of the nature of (some kinds of) work during this period of rapid economic and technological change.

The book provides a rich analysis of routines and change in four case-study companies. Explanations of the behaviour of the firms are developed and provide new concepts which can help in the understanding of other firms in different circumstances. The work emphasises empirical evidence, rule-based action and the need to recognise the historical, social and interpretative contexts.

The primary contribution of the research is the development of a broader concept of routine, in particular the identification of routine practices at a strategic level and the demonstration that such practices can incorporate change. In addition, the analysis examines the role of technology in economic change: the extent to which technology can be seen as an agent in its own right. I believe the evidence here confirms that an institutional approach to the understanding of firms' behaviour is fruitful and can add to the current repertoire of approaches in economics.

I am deeply indebted to Maureen Mackintosh and Matthew Jones who have provided intellectual stimulation, thoughtful criticism, inspiration and encouragement which forced me to think for myself and to apply their rigorous standards to my work. I have learned a great deal from them about the nature of research and how to carry it out. I have also benefited from attendance at the Realist Workshop at King's College, Cambridge. Participation in the workshop requires concentration and clarity of thought and I have gained valuable insights from comments on my work, particularly from Steve Fleetwood whose encouragement and enthusiasm first drew my attention to the lively debates at King's on Monday evenings. The Information Systems Forum at the Judge Institute of Management Studies have been generous in giving ideas and support to someone only tangentially connected with their interests. I have also been fortunate to work with a lively and supportive group of colleagues in the Economics Discipline at the Open University. Their interest and comment have been a source of real learning both explicitly and in the day-to-day round of work in the university. I thank them all. As always, however, all errors and omissions are, unfortunately, entirely my own work.

Special thanks go to the companies participating in the research and particularly to the staff who agreed to be questioned about their work and generously gave their time. Thank you also to Tony Richardson for high-quality tape transcription, Avis Lexton for putting the text in order and to Tricia Smoothy for invaluable support in actually getting the research work done.

Finally, a big thank you and unrepayable debt to Ginny, Anna and Tom, who have left me alone when I needed to be left alone, and listened or encouraged when I needed those things and without whom I could not have given the time and energy needed to carry on.

Neil Costello
February 2000

# 1   Introduction

How do firms live with change? What makes some firms thrive and others struggle when faced with changing environments? And how important is high technology in driving or facilitating change? These were the questions which motivated this project. Other questions followed. Is change revolutionary or evolutionary? Are firms subject to transformations arising from the new technologies or do they absorb them incrementally? How much is planned or anticipated, how much occurs by stealth?

'Habits gradually change the face of one's life as time changes one's physical face; and one does not know it' (Woolf 1926). This eloquent expression neatly sets out many of the issues here. The lives in this case are those of four high-technology firms. Their physical faces and organisational practices changed during the period in which they were studied. They lived in changing circumstances with the pressures of rapidly changing technologies. In some respects they appeared to be tossed about in worlds undergoing major shifts but the habits and routines they had adopted gave them a certainty and predictability in their lives so that they changed but did not always know it. The study of the firms presented here examines the manner in which some of those changes took place and searches for causal factors and explanations.

The motivation for the project was to understand how firms, as particular economic institutions, lived through and experienced change. Routine behaviour provides the framework through which the analysis of the firms is undertaken. The idea of routines has become increasingly influential. Here we use that concept exhaustively. Can we observe routines? What do they do? The project pushes the notion of routine very hard. To use an experimental analogy, it is tested to destruction. As the term is used here, a routine is the taken-for-granted way of working which is usually unchallenged. Routines are 'the way we do things around here'. They are the recurrent practices and regularities in company behaviour. They are not fixed but changes in them are slower than changes elsewhere in the organisation. As subsequent chapters go on to show, the concept is valuable in understanding change in organisations.

## Objectives

There are three broad objectives. Firstly, the book is attempting to contribute to a richer social ontology through which to understand firms' behaviour. The static analysis of conventional neo-classical economics is unable to deal with processes of change and organisational behaviour. But in alternative approaches to analysing firms, such as those adopted by the institutional schools, routine behaviour is often taken for granted as a low-level activity while frequently being seen as inappropriate at policy-making or strategic levels. In the management literature, conversely, routine behaviour is frequently undifferentiated from organisational culture so that the term 'culture' becomes a catch-all for everything the firm does. The conceptual framework can be enriched by distinguishing, as we do here, between routines and cultures and by investigating, empirically and theoretically, the nature of routines themselves.

The second objective is to consider the relationship between change and routines. An important consideration is to study the individual practices and structures which enable the firm to identify and engage with change. Routines form part of the practices and are implicated in changes in the structures. As will be shown, routines do not necessarily inhibit change; they can drive people to change and change can be incorporated as part of the routine.

The third objective relates to the implications of technology for organisational change. Electronic technologies have become commonplace and are seen in some interpretations to be the driving force behind change in firms, if not in the whole economy. The determinist view of technology is that it drives change: because of the technology, change will take place. Frequently an unquestioning assumption is made that change is dependent upon technological breakthroughs and that its direction, if not its precise form, is inevitable. This project challenges such claims and investigates directly the ways in which electronic technology is implicated in organisational change.

Additionally, technology has important practical implications for the project. In order to study routine behaviour it is necessary to choose a terrain in which change is taking place. Clearly an analysis of routines and change in a calm, unruffled area of the economy, if such an area exists, would produce quite different opportunities from the study of an area undergoing turbulent change. Unruffled calm does not allow us to differentiate between those things which are unchanging because the world is constant and those things which are unchanging because they are relatively enduring regularities. The firms studied were chosen, therefore, because in different ways they were engaged with advanced technologies. The hypothesis was that firms working in high-technology areas would be likely to face elements of rapid change in at least some aspects of their work and that this would facilitate the study of organisational routines. Technology was thus a vehicle through which to assure change as well as an object of study in its own right.

## Routines, structure, agency and culture

Routine as a concept has been used for many years in the economics and economics-related literature but Nelson and Winter's classic text (Nelson and Winter 1982) developed the idea in a more systematic and theorised way than had been attempted hitherto. It distinguished three classes of routine: short-run routines reflecting the operating characteristics of the firm; routines which determine decisions about the firm's capital stock; and routines which modify operating characteristics. Routine-guided processes were modelled as 'searches'. Nelson and Winter developed these conceptual categories primarily as modelling tools. They were grounded in empirical work but were not developed as empirical categories.

This project takes those categories and examines the behaviour of four companies empirically. Two levels of routine are identified: operational routines and strategic routines. They cannot be completely separated but broadly operational routines are concerned with the continuing reproduction of the firm. They enable it to continue doing the things it does. Strategic routines are concerned with the place of the firm in its wider environment. The research establishes that the firms did not use some kind of maximising or satisficing rule but responded to change following taken-for-granted patterns of behaviour. Such patterns, defined as routines, were not determining but indicate an evolutionary and path-dependent world. In some cases the routines which evolved enabled the firms to incorporate change into their taken-for-granted patterns of working. Far from inhibiting responses in novel circumstances, the routines themselves have made it possible for the firms to cope better with change.

Subsequent chapters identify the mechanisms through which these routines were sustained and reproduced. To undertake this requires an examination of the relationship between structures (both within the organisation and from the wider economy) and agents. Structures are defined here as the formal and informal rules which govern behaviour and the relationships which depend upon them. They may be written down as procedures or understood on the basis of widely accepted norms based in more general rules. They are reflected in the reporting hierarchies of the companies studied and partly define the position and status of staff and others connected with the company.

Agency is seen as human agency. However, in the process of the research upon which this book is based, it became clear that agency could not be separated entirely from the materiality of the world. The extent to which technology can be conceived as an agent in its own right is addressed here. In some cases information technology is deeply constitutive of the firm and recognition of this was influential in the realist methodological position which is adopted in subsequent analysis.

Culture is used as a category which reflects and acknowledges the 'cultural turn' in the human and social sciences (Hall 1997). This approach emphasises the importance of meaning for the definition of culture. Culture, in this interpretation, is not a set of things, such as paintings or novels, but is to do with the production and exchange of meanings. If a group of people belong to the same culture then they share a set of meanings about the world: they interpret the world in the same way. That sharing is not complete. Within the same culture there will be subtly different interpretations and different ways of representing meanings, and those differences are implicated in the changes in culture and the different practices which represent it. The question of meaning arises in all aspects of the social world. Artefacts have symbolic meaning; for example, in one of the case-study companies, the company's building represents much more than simply a place of work, and meanings are communicated through such things as gesture and dress, as well as importantly through language. Hall describes language as 'one of the privileged "media" through which meaning is produced and circulated' (Hall 1997, p. 4).

Routines are practices which are situated in a particular set of meanings. This project uncovers those practices by interpreting the actions of human agents in different structural positions in the four companies studied. The interpretations draw heavily on the participants' own interpretations of what they were doing and why. The primary research evidence is the language of the participants and the meanings they place upon the events, symbols and relationships they describe.

An interpretative approach is therefore central to the project. Underlying that approach is a recursive explanatory framework. Routines and technology appear as both explanation and explanandum. The explanation involves self-reflection on behalf of the agents observed. They are not seen as automatons. Self-reflection is part of the recursive framework: human agency involves reflection on what has gone before so that behaviour and structures are modified, frequently in a non-deliberate way, in the light of experience. Giddens calls this the double hermeneutic (Jones 1998a). The explanatory framework is therefore also path-dependent. The place from which an individual or company starts is partly affected by where it has come from, and this in turn restricts the directions in which it can go.

The categories which are discussed and developed provide explanations for the behaviour of the four firms studied and provide new building blocks for the understanding of other firms in different circumstances. Specifically, routines are used as an explanatory concept at strategic levels in organisations; they are also used to explain the ways in which firms engage with and incorporate change into their practices. This has not been done before. Subsequent chapters argue that the use of the concept in this way adds significantly to our understanding of the behaviour of the firms studied.

Within this framework the relationship between economic and technological categories is explored. This contributes an understanding of the extent to which technology is constitutive of firms, that is the extent to which it has the power to enact or establish firms. In the case studies reported here, electronic technologies possess some of the characteristics of change agents through sharing information, providing the potential for new forms of control and, through these characteristics, changing the perceptions individuals have about themselves and their roles. Zuboff (1988) would describe these features as informating. Technological products also contribute to the interpretations firms make of their external environments. Technology can be seen to both enable and constrain organisational behaviour but, as will be shown, it cannot be adequately analysed outside the social system of which it is a part.

## The research process

The research questions produced a logic in the choice of fieldwork. Analysing change meant that the study should be longitudinal. A number of firms which had clear differences needed to be investigated, so that routine practices had potential contrasts. The firms also had to be concerned with new electronic technologies, but in different ways, so that the implications of technology for change could be better understood.

Four organisations were eventually selected with characteristics which met these criteria. All were in the same local economy, greater Cambridge, in order to minimise distinctions arising from spatial differences and to engage with a region which, by UK standards, was undergoing rapid change. They were all small to medium sized. At the beginning of the research period all employed over fifty but fewer than 100 people. All were involved in working with the new technologies in some way.

Two of the organisations chosen were in the new-technology business. One was the East of England branch of a major, well-established computer manufacturer. It manufactured, serviced and provided consultancy services in computing, and so was directly involved in the development of ideas about technology as well as in producing the artefacts themselves. The second was a supplier of advanced software and internet services. It was at the forefront of what was then conceived by many commentators to be the most dynamic and uncertain part of the economy. It was entirely focused around changes in the new technologies.

Studying only organisations like these, however, would have focused too much upon the experiences of a particular sector. It was important to look also at organisations which use technology, perhaps in a very sophisticated way, but whose business is not entirely concentrated in the technology, in order to see if contrasts exist in their experience and whether routines, change and technology interact in similar ways for firms outside and inside the sector. Two further organisations were approached, therefore, in order to broaden the perspective. The first of these is a publisher which publishes through new technologies. While it is a technologically sophisticated company, its focus of attention and its strong sense of its own image are those of a publisher, not of a high-tech firm.

The last organisation participating in the research was less technologically sophisticated. At the design stage in the research it seemed desirable to study an organisation which was undergoing major technical change but which had not used technologies in a significant way before. The fourth company was a small, successful trust producing distance-learning materials (and therefore with some similarities to the publisher) which had used some computing facilities but was in the process of implementing a major information systems redevelopment. It was therefore possible to study the ways in which the new technologies were perceived and how they fitted into previously acquired routines as the changes associated with the implementation of the new technologies were taking place.

Fieldwork was carried out in all four organisations over a two-and-a-half-year period. It became clear early on that routine behaviour existed in each institution. The research then focused on the features of the organisations upon which routine behaviour was based. Where did routines arise, how were they reproduced and transformed, and what part did the new technologies play in this?

The research process concentrated on recovering agents' meanings but recognised that there is an irreducible materiality to the world. That materiality exists independently of an agent's knowledge and interpretation. Recovering agents' meanings only is therefore insufficient. The meanings must be assessed and examined against other explanations of the world. The approach here is to use grounded theory which searches for patterns in the agents' explanations and descriptions. The patterns identified are based in metaphors and pre-existing explanatory frameworks held by the researcher. They are built by comparing different interpretations in interview or documentary evidence and placing them alongside other plausible explanations. Peirce, one of the founding philosophers of institutional economics, characterised this method as 'abduction' (Mirowski 1987), in which the researcher, in forming an explanatory hypothesis, tries to match new events with explanations already understood. In this process the pre-existing explanations can change.

# 2 The theoretical framework

The more we can learn about the way in which firms actually behave, the more we will be able to understand the laws of evolutionary development governing larger systems that involve many interacting firms in particular selection environments.

(Nelson and Winter 1982, p. 410)

In order to learn we must impose patterns on phenomena; that is the only way in which we can make sense of them. We may, of course, be making sense of what is not really sensible, especially when the subject of our study is human behaviour. So the assumption that human behaviour is based on reason, although a much weaker version of the standard economic assumption, can reasonably be criticised for assuming too much rationality.

(Loasby 1991, p. viii)

Undoubtedly part of the problem reflects the still primitive state of our ability to work with cultural evolutionary theories. In this particular case I am sure it also stems from an overly broad and vague concept of the variable in question – institutions – which is defined so as to cover an extraordinarily diverse set of things. Before we make more headway in understanding how 'institutions' evolve we may have to unpack and drastically disaggregate the concept.

(Nelson 1995, p. 84)

Where should the analysis go next? . . . First, for instance, it is necessary to examine the particular origins of those habits and rules. Second, the ways in which new rules and habits are created and displace others have to be addressed. Third, the criteria of efficacy have to be considered, including cases where habits or rules are more useful in some contexts rather than others, or may be advantageous for groups but not for individuals, or vice versa. Fourth, the mechanisms by which habits and rules build up to social routines and institutions have to be analysed, as well as the feedback loop by which institutions help in turn to reinforce particular habits and rules.

(Hodgson 1997, p. 681)

## Introduction

The four quotations above, drawn from the work of economists who are critical of mainstream ideas, show how little our knowledge of the detailed working practices of firms has increased since Nelson and Winter's classic text of 1982. In 1982 Nelson and Winter argued that we need to learn more about the way firms actually behave. This would be the beginning of a new approach to theorising and a search for 'laws of evolutionary development'. Loasby argued for imposing patterns on phenomena that recognised earlier moves away from models of rational maximisation. His approach implied the need to look for different decision methods in firms through 'making sense of what is not really sensible'. This would include imposing patterns, which are described here as routines, of which, as Loasby describes, no body of knowledge existed in 1991. In 1995, thirteen years after Nelson made his original request (with Sidney Winter), he continued to bemoan the vague generalisations in use and argued for the unpacking and disaggregation of concepts. Hodgson sets out a research agenda: to examine the origins of habits and rules, to look at the creation and displacement of habits, to consider the efficacy of rules or habits, and to consider the mechanisms by which habits and rules build up to social routines and institutions.

It is time to examine such concepts in practice and in detail. Concepts such as routines, rules and habits are being used increasingly, and 'institutional' is now a descriptor for a major research orientation in economics. This project falls within the institutional framework. It is concerned with understanding and theorising economic institutions but, as will be picked up later in this chapter, the framework used here is closer to that of the 'old' institutionalists than the 'new'.

## Culture and the value of interpretative approaches

Developments in evolutionary economics link closely to ideas coming from organisational theory and sociological theory. The notion of a *routine* which is a major analytical tool within evolutionary economics can be approached from the perspective of organisational culture but routines are different from culture. Routine is used as an abstract, general, theoretical tool by evolutionary economists and has been valuable for modelling the behaviour of firms. Computer simulation of firms' behaviour has been developed from such models. Empirical work in evolutionary economics, however, has concentrated on the material world, the traditional focus of economics, and has considered almost exclusively technical change and innovation (for example, Leydesdorff and Van Den Besselaar 1994; Lundvall 1992; Andersen 1994). This was also the original motivation for Nelson (Nelson and Winter 1982, p. vii).

In contrast, the research here focuses on the human and social side of the organisation and studies the different perceptions of individuals in different structural positions within the four case-study companies. In looking through their eyes the objective is to find the key events, beliefs and symbols which they use in making sense of the organisation and in particular the way in which economic decisions are formed. There is a need to look at events and processes as the subjects perceive them (Boland 1985; Johnson 1987; Walsham 1993). The organisation is the output of the processes under study and is not some kind of entity which can have opinions of its own or carry out its own actions. In order to understand the organisation, therefore, it is crucial to gain an understanding of the perspectives of the members of the organisation. Their combined understandings are essentially the organisation's understandings. Such a focus is an interpretative one.

An interpretative approach emphasises the impossibility of finding completely objective knowledge. We cannot strip away all prejudice, or pre-existing theory; we interpret the world through experience. Understanding the behaviour of firms then requires a reading of the interpretations of those involved in the firm, but a recognition that the reading we are undertaking is also being interpreted through our own particular set of theoretical lenses. *Reading* is a useful metaphor in this context. Frequently it is literally the activity undertaken – reading interview notes or the accounts given by other people – and it emphasises the active participation of the reader in the interpretation of the events under consideration.

An interpretative, cultural and political approach to organisational change is fruitful because it relates to the meanings perceived by actors within any organisation about how that organisation works and to the importance of its culture and the symbols which are used as part of that culture (Johnson 1987, 1988). However, it does not imply that phenomena can only exist within the mind of the agent. Features of the world can be independent of agents. Lawson (1999) would call them intransitive, that is 'irreducible, for the most part, to our knowledge of them' but we make sense of them through our interpretations and meanings, our culture.

Culture is not an exogenous feature of the organisation's world. It is reproduced and transformed by the social and structural relationships inside and outside the organisation. Culture as used here refers to the production and circulation of meaning (Du Gay *et al.* 1997). Culture is about shared meanings in which representation through language is a central process in the way meanings are produced (Hall 1997).

Cultures in organisations are frequently complex. Johnson, for example, argues that organisations have a complex *cultural web* centred on a paradigm. Surrounding the paradigm are the features of organisational life, the web, which overlap and interrelate so that

> organizational environments take on meaning, not independent of the organizational context, but within it; specifically that organizations are likely to have more or less homogeneous ideologies, the constraints of which provide meaning and legitimize organizational action . . . there are

likely to exist a core set of such beliefs that are taken for granted, endure over time and account for the way the business competes . . . these beliefs . . . are embedded in the cultural and political web of organizational life. They are not remote from the day-to-day lives of managers but part of them.

(Johnson 1987, p. 230)

The cultural web is thus effectively an organisational ideology. Johnson includes 'rituals and routines' as an element of the web. These are the only *practices* in the web. In this project they are extracted as a separate category linked to meanings but different from culture.

Culture does not exist only at the organisational level. 'Recipes' also exist at the industry level (Grinyer and Spender 1979; Spender 1989). Culture is connected to industry-wide recipes as well as the local culture of the organisation. Similar firms in different industries can adopt different cultural meanings which are similar to those adopted by other firms in their industry. Spender defines recipes as 'the industry's pattern of managerial belief' (Grinyer and Spender 1979, p. 116). These ideas have a largely descriptive, static quality.

The external world is interpreted by individuals inside the organisation and is not separate from the internal world though it is frequently considered as such within organisations. This sense of internal and external context is seen as a crucial feature by some (Pettigrew 1985), though Morgan (1989, 1986), following Weick (1979), supports the argument that the distinction between the organisation and the outside world is not so sharp as is usually assumed. For interpretative organisation theorists the environment is 'an *enacted* or socially constructed domain that is as much the consequence of the language, ideas, and concepts through which people attempt to make sense of the wider world as it is of the "reality" to which these social constructions relate' (Morgan 1989, p. 91). Thus organisations are themselves players in, and have some control over, their own environments. Furthermore, the environments are themselves interpreted through constructions which are derived to a considerable extent within the organisation. This approach is important in understanding the companies analysed here. It draws attention to the constructions made by companies and the way these are implicated in company practices.

Shared meanings and beliefs need to be constructed and maintained (Pfeffer 1981). The role of ritual and ceremony can be important but shared meanings are not embodied solely in the more formal aspects of organisational life. Mundane behaviours too are important transmitters of organisational norms and beliefs (Nadler and Tushman 1990). Shared meanings are not seen in this project as elaborate, uniform, monolithic structures but as stocks of knowledge which are held in common and drawn upon (Boland 1996). In interpreting the routine practices of companies, we are studying the ways in which different actors draw upon that stock of knowledge and how, in doing so, they become accountable to themselves and to others.

## Change processes

Routines are relatively enduring features of firms and need, therefore, to be set in the context of change. This project is focused on change. It concentrates on the dynamic aspects of firms' behaviour not just as a comparative, static series of pictures, but to try to discover the ways in which firms generate and deal with change. Time is not simply elapsed time but has a chronology and a history. Events are essentially historical and influence activities in a path-dependent way. This focus has a long history and was a problem with which Marshall struggled in writing his *Principles* (Thomas 1991). Path dependency also has very clear practical implications, frequently at a technical level, well demonstrated by the history of the QWERTY keyboard (David 1985).

Accepting path dependency and the possibility of an evolutionary approach requires that antecedent conditions must be considered in analysing decisions. History should be considered not as an event in the past but as part of current conditions and understandings. Context is about the interpretations and meanings which actors make when undertaking the process of decision making (Pettigrew 1985). The strategic agenda has to be set in context. It is necessary to consider how items get onto the strategic agenda, and to look at agenda *building* rather than agenda *setting* (Dutton 1988). We should also distinguish between the objective conditions faced by a firm, the cognitive arena, and the network of potential and actual collaborators (Child and Smith 1987). Different outcomes will arise depending upon the balance of these features. Child and Smith argue that the legacy of a firm's history bears heavily on its ability to effect transformation. The firm is encumbered by founding ideologies, sedimented structures and distinctive competences which may no longer be suited to its competitive environment. Routines themselves become sedimented and we must consider the mechanisms or processes which enable (some) organisations to manage this.

History must be taken seriously. Chandler (1962, 1977, 1988) is pre-eminent among those who have done so. He insisted that a meaningful analysis required accurate knowledge of a firm's organisation and admin-istrative history because organisational structure was intimately related to the ways in which the enterprise had expanded. His work gives an excellent justification for a historically specific case-study approach. He argued that:

> The underlying argument . . . is that the impact of changing technologies and markets on economies of scale and scope and on transactions costs – and on the organization created to exploit those economies – provides the most satisfactory answer to the basic historical questions of why the large multi-unit industrial enterprise came when it did, where it did, and in the way that it did.
>
> (Chandler 1988, reprinted in McCraw 1988, p. 476)

Chandler's analysis was influenced significantly by a Parsonian structural-functional approach, however. He believed that social science was value-neutral and he had little interest in intellectual history (McCraw 1988). He imposed his own framework on his detailed and thorough case studies, and emphasised the importance of rational planning. Chandler's emphasis on rationality and the value-neutrality of the social sciences fails to recognise the social construction of institutions and processes. There is an increasing recognition that models of strategic change which emphasise rational planning are no longer adequate (Child 1972; Mintzberg 1990; Mintzberg and Quinn 1991). Mintzberg has continually argued there is no one best way of managing. In developing a number of 'configurations' of organisations, his work is valuable in emphasising the emerging, socially constructed nature of strategy formation and in setting up models or metaphors through which to consider other organisations.

A number of writers, including Mintzberg, have also noted that organisational change tends to continue in small steady steps punctuated by larger quantum jumps (Tushman and Romanelli 1985; Mintzberg and Waters 1982; Miller 1982; Miller and Friesen 1980). The claim has been disputed (Quinn 1980) but there has been some convergence of views (Mintzberg and Quinn 1991). Overlaying the claim was the earlier observation (Mintzberg 1978) that the behaviour of senior managers in organisations frequently did not resemble the models found in many management textbooks, in particular those which emphasise rational planning and decision making such as Chandler. Managers' actual behaviour involved more responsive and immediate judgements and rested upon a sensitivity to the environment rather than the firm implementation of previously devised plans. Mintzberg argued that strategy emerged from patterns of actions rather than from rational planning. Patterns of actions in Mintzberg's sense have a strong similarity to the idea of routines presented in evolutionary economics. Mintzberg does not dwell upon the manner in which actions become routinised but his configurations are in many respects examples of particular combinations of routinised actions which suit particular environments.

## Routines

The search for routines, and how they are constituted and changed, is the central focus of this project. The idea of a routine is at least implicit in the organisation theory and sociological theory which have been reviewed so far. The notion of sedimented structures and routinised behaviour is common to the literature on organisational culture. Patterns of actions are seen as important in the literature on strategic change processes. These can be considered to be routines at a strategic level. The idea of organisational learning implies a set of relatively routine behaviours which promotes benign responses to stimuli, and an equivalent set of routines which inhibits learning.

In economics the term routine is used in a conceptually similar way but plays a more significant role in the development of theory than has been the case so far in the other social sciences. Douglas (1987) describes economists as the strong theoreticians in the social sciences, typically concerned with ideas about the *material* world, using applied science and rigorous theory which is measurable. Empirical work has concentrated on the more directly measurable features of routine behaviour and their outcomes, such as those connected with technical change. This study is attempting to look at features which bring human dimensions of routines to the forefront, and the approach which is taken to routine behaviour in evolutionary economics can be of considerable help.

Evolutionary economics challenges the (comparative) static models which have driven the discipline in the past and focuses on processes of change using a biological analogy. Nelson and Winter (1982) conceive of routines as the genes of an organisation. Routines are acquired by organisations in the same manner as skills are acquired by individuals. Organisations become skilful and consequently do not take decisions but simply undertake processes of action. The more successful organisations survive, as in natural selection, through the interplay of their genes and their environments. The biological metaphor is powerful, partly because it has an applied focus and yet is sufficiently abstract to generate explanations across many organisations.

There is now a considerable literature arguing that economic analysis should move to an evolutionary approach, as well as a number of classic articles and texts which are consistent with such a move (for example, Alchian 1950; Penrose 1952; Boulding 1981; Matthews 1984). As yet there is no empirical work which attempts to analyse the constitutive processes involved in the creation and reproduction of routines in firms and that is the area in which this project makes its contribution.

Ideas about routines or rule-bound behaviour are used increasingly in attempts to understand the behaviour of organisations. Routines are seen, at a common-sense level, as predetermined, unchanging patterns of behaviour exemplified by the kind of activities which workers carry out on production lines. Thus passing partially completed sets of components from one worker to another, each adding a small contribution to the finished product, is the common image of routine. This image has been used metaphorically in the literature but it is not entirely appropriate. For example, Cohen and Bacadayan (1994) concentrate on the individual and psychological dimension of routine behaviour. Their definition of routines as 'patterned sequences of learned behavior involving multiple actors who are linked by relations of communication and/or authority' (Cohen and Bacadayan 1994, p. 555) is appropriate only for a narrow range of behaviour untypical of many of the practices of organisations.

Routines are a response to the quantity and complexity of information faced by firms. They are not a conscious maximising choice, but are where the firm's organisational knowledge is stored. They are a source of difference among firms which, through a selection mechanism, drives the evolutionary process.

This form of analysis was prefigured by the distinction between resources and services (that resources render) (Penrose 1959). It is the services which are a function of the experience and accumulated knowledge of the firm and which are the source of the firm's distinctiveness. Dynamic capabilities (Teece *et al.* 1990) which are sticky, i.e. not easily changed, acquired or passed on, build on this distinction. Such capabilities are inherited from and constrained by the past and from them the firm develops a set of skills, assets and routines which become its core competences. Such soft assets cannot be traded and so must be built, often over decades.

Routines are the regular and predictable aspects of firms' behaviour. Modelling the firm requires modelling routines and observing how they change over time. In practice, rules and procedures must not be too complicated because of the bounded rationality of people in firms, but the world possesses an excess of information (excessive, that is, in relation to the ability of individuals or organisations to process it). Recognition of this point was the focus of the seminal work of March and Simon (1958). They define routinised behaviour:

> Activity (individual or organizational) can usually be traced back to an environmental stimulus of some sort. . . . The responses to stimuli are of various kinds. At one extreme, a stimulus evokes a response – sometimes very elaborate – that has been developed and learned at some previous time as an appropriate response for a stimulus of this class. This is the 'routinized' end of the continuum, where a stimulus calls forth a performance program almost instantaneously.
>
> (March and Simon 1958, p. 139)

March and Simon's definition is important in insisting that firm differences matter and therefore that analysis of firms as unique and important institutions in themselves is an important activity. March and Simon discuss a continuum which describes behaviour as moving from completely routinised to 'problem solving', the point where new performance programmes may be constructed. A set of activities are regarded as routinised 'to the degree that choice has been simplified by the development of a fixed response to defined stimuli' (March and Simon 1958, p. 142). The research reported here will show that this is too stark a distinction in practice.

The variation and openness of routines are often missed. Grant (1991), for example, follows Nelson and Winter (1982), and defines routines as 'regular and predictable patterns of activity which are made up of a sequence of co-ordinated actions by individuals'. This misses the subtlety of Nelson's later analysis (Nelson 1995) and fails to recognise that routines can be *un*predictable in the sense of determining any specific response, while nevertheless retaining a recognisable pattern or approach. It emphasises *predictability* over *repetition*. A routine need not be a clear sequence of co-ordinated actions. Indeed it would not normally be that. Skills and the routines based partly upon them are likely to be complex.

Grant goes on to argue that there is no predetermined functional relationship between resources and capabilities. He argues that style, values, tradition and leadership are all critical to capabilities. Given these claims, routines could not, in their manifestations in the world, be regular and predictable patterns of activity. However, since routines do produce automatic responses and are based on skills, which are themselves a form of tacit knowledge (Nelson and Winter 1982), it is important, he claims, to recognise that their efficiency in dealing with day-to-day affairs can be constraining when novel situations arise. Experience is a valuable asset, though, for Grant, it becomes less important when change is rapid. But the experience of coping with change can itself become a skill and build routine forms of behaviour. The analysis of Unipalm-Pipex and Chadwyck-Healey in Chapters 6 and 7 shows how change can be incorporated into routine behaviour so that, in contradiction to Grant's claims, routines can be helpful in novel situations.

One of the few genuinely empirical approaches to routines (Pentland and Rueter 1994) addresses some of these issues but, like the work of Cohen and Bacadayan (1994), who are the other significant contributors to empirical work in this field, then concentrates on relatively low-level activities. Pentland and Rueter point to the critical position routines occupy in the analysis of organisations and argue that they are 'complex patterns of social action'. Confusion arises, they claim, because routines are thought of as a 'capacity' rather than as a pattern of action (Pentland and Rueter 1994, p. 484). In studying routines, Pentland and Rueter emphasise the sequential structure of the patterns and build on earlier work by Pentland in adopting what they call a grammatical model.

The grammatical model has interesting parallels with the work in this project. It emphasises the nature of routines as patterns of action and, in doing so, shows how the grammar (using a metaphor of grammar and language) can produce different manifestations of the same basic structure. The grammar does not specify an outcome but defines the possibilities from which organisational members can accomplish particular sequences of action. Both structure and agency are therefore acknowledged by this model, as they are here.

Because routines are complex patterns of action, they cannot be seen as automatic responses. Rather they should be understood as effortful accomplishments. Routine work is not mindless or automatic:

> The regular or routine features of encounters, in time as well as in space, represent institutionalized features of social systems. Routine is founded in tradition, custom or habit, but it is a major error to suppose that these phenomena need no explanation, that they are simply repetitive forms of behaviour carried out 'mindlessly'. On the contrary, as Goffman (together with ethnomethodology) has helped to demonstrate, the routinized character of most social activity is something that has to be 'worked at' continually by those who sustain it in their day-to-day conduct.
>
> (Giddens 1984, p. 86, quoted in Pentland and Rueter 1994, p. 488)

Routines, then, are organisational practices and there is value in decomposing economic systems and in trying to discover insights about their component parts. The concept can be used in the analysis of patterns of behaviour which require some thought even though the behaviour may be in some senses automatic. A stimulus may prompt the firm towards a particular form of analysis or other form of complex behaviour. The behaviour is routine in the sense that those sorts of problems are always analysed in that way but it is not simple. Here the concept is used to refer to the *taken-for-grantedness* of behaviour and it has close links with the concept of institution.

One of the fundamental premises of institutional economics is that the determination of whatever allocation occurs in society arises from the organisational structure – the institutions – which change through 'non-deliberative' (habit and custom) and 'deliberative' (legal) modes (Samuels 1995). Nelson defines institutions as 'a complex of socially learned and shared values, norms, beliefs, meanings, symbols, customs, and standards that delineate the range of expected and accepted behaviour in a particular context' (Nelson 1995, p. 80). This is consistent with the broader approach to institutions which comes out of anthropology. Douglas (1987), for example, sees an institution as a convention, which she defines, following Lewis (1968), as arising 'when all parties have a common interest in there being a rule to ensure co-ordination, none has conflicting interest, and none will deviate lest the desired co-ordination is lost' (Douglas 1987, p. 46). Such conventions are self-policing and Douglas refers to an institution as a *legitimised* social grouping. The legitimacy can come from many places but specifically excludes any purely instrumental or practical arrangement that is recognised as such. Nelson's 'expected and accepted' behaviour is close to Douglas's legitimacy. For the new (economic) institutionalists, institutions are the rules of the game (North 1990) where the emphasis is on the self-reinforcing nature of institutionalised behaviour, a form of self-policing convention. Path dependency is an important implication of these definitions.

In this project established practices at a societal level – institutions – are distinguished from those within the organisation – routines. *Routines are defined here as established, significant, sanctioned and recurrent practices within organisations.* The research searches for such practices in the four organisations studied. It seeks to explain the ways in which such practices are created and reproduced. Those explanations are set in a recursive framework linking structure, agency and technology.

The research uncovers routinised practices. Not all behaviour is routinised but the argument is that a great deal of behaviour can be understood in this way; therefore to make a distinction between completely routinised and problem solving behaviour is not adequate. It is important to take into account the routine strategic practices which we can observe in firms and to account for the ways in which operational and strategic practices interact.

## The interrelationship of structure and agency

An understanding of routine behaviour in firms thus requires an analytical framework which can bring together cultural, institutional and individual levels of analysis. In economics the development of theory of this kind has been taken on by the institutional schools.

Institutional economists have taken a greater interest in methodology than more orthodox schools. This is unsurprising, given the need for any heterodox group to be clear about the distinctiveness of its analysis. The 'old institutionalists', building on the work of Peirce, Commons and Veblen in particular, have been largely an American phenomenon, though the 'Methodenstreit' of the German historical schools has also been influential (Becker 1998). They have largely rejected comparative static modelling based on closed systems and *ceteris paribus* assumptions, and concentrate on the analysis of institutional power and the relationship between economic structures and human agency. The 'new institutionalists' come from a more orthodox background and are less concerned with change processes and the relationships between agents and structure. The analysis in this project falls into the area of interest of old institutionalism but uses a theoretical framework which has not been well developed there.

Institutionalism has been relatively eclectic, so incompatible modes of analysis can fall within its remit. It is a broad church. However, it is generally understood to have a number of common features (Samuels 1995) and they form the framing assumptions of the analysis set out here. The emphasis is on social and economic evolution and it is argued that institutions are relatively enduring and socially constructed. Furthermore, the market economy is itself a system of social control which operates through institutions and is formed by institutions. Thus the idea of a self-subsistent individualism operating through a mechanical mode of theorising in the quest for static, determinate, optimum equilibrium results is seen as an inadequate framework in which to understand the behaviour of economic institutions.

Technology is perceived as a major force in the transformation of economic systems but 'it is human activity mediated through technology that determines what is a resource, its relative scarcity and efficiency' (Samuels 1995, p. 573).

For institutionalists, culture has a dual role both transcendentally in the formation of structures and identities, and as an artefact of the continuous interdependence among individuals and sub-groups. The relationship between structure and agency which is implicit in this institutional formulation lies at the core of the analysis.

### Structuration

Structuration theory (Giddens 1982) emphasises the recursive relationship between agency and structure and is valuable in forcing an acknowledgement of the key processes involved. Giddens describes structures as traces in the mind which exist only through the action of humans. His objective is to break out of the unsatisfactory *dualism*, as he sees it, of agency and structure, and instead to focus upon a *duality* of structure as the essential recursiveness of social life. Structure is both the medium and the outcome of the reproduction of social practices. In this perspective social science is irretrievably hermeneutic.

The concept of structuration is subtle. It arises from the long philosophical debate which sees agency or action by individuals to some extent in opposition to ideas of social structure or social system. It has parallels with debates about the extent to which individuals have free will or how far individual behaviour is predetermined or systematically constrained in some way. Giddens spends many pages discussing the relationship between the concepts of action and structure. He defines structure as 'recursively organised rules and resources' while structuration is seen as 'conditions governing the continuity or transformation of structures, and therefore the reproduction of systems' (Giddens 1982, p. 35). What this comes down to is the need to conceptualise a structure which is constantly maintained and changed by the actions of the individuals who operate within it. The individuals are not necessarily aware that their actions are both confirming the existence of the structure and changing it. In turn the structure, as a set of rules and resources, is independent of the individual, whom it constrains and empowers, but it exists only in the instantiation of human action. The arguments link to the notion of *enactment* set out by Weick (1979), in the sense that the structural properties of social systems are, for Giddens, the outcome of the practices which constitute those systems. He is arguing that to understand organisations we must consider their structures and the processes in which individuals engage which legitimise and define those structures. Garnsey (1992) talks about these as the constitutive processes which exist in dynamic social systems.

Within structuration theory social rules are distinguished in similar ways to those of the institutionalists. Giddens talks of rules of social life and formulated rules. The latter are codified interpretations of rules existing in games or bureaucracies rather than rules as such. This is more complex but broadly matches the deliberative and non-deliberative distinction of the institutionalists. For Giddens, action is a 'continuous flow of conduct' (Giddens 1979, p. 55) which is reflexively monitored. Actors are assumed not to have conscious goals but to follow rules. They are knowledgeable and know how to play according to the rule. Rules thus generate practices.

For Giddens, however, routine has components additional to those already discussed. Routines are integral to the continuity of the personality of the agent and to the institutions of society. They give the agent ontological security. Ontological security is important for Giddens. It is the system of inner security which is protected by social devices such as tact or saving face and maintained in a 'fundamental way by the very predictability of routine' (Giddens 1984, p. 51). This gives additional and helpful insights and can be incorporated into accounts which do not rely on the full apparatus of structuration theory. Bounded rationality contributes to insecurity and it is partly the search for ontological security which pushes organisations into developing routine forms of behaviour.

Giddens is one of the most influential living sociologists but his work is controversial. He does not claim that structuration theory should, or indeed can, provide a framework for social research, explaining that it is 'relatively autonomous in respect of research' (Giddens 1990). His claims are ambitious, however:

> I have never thought of structuration theory as providing a concrete research programme in the social sciences. . . . It is an attempt to work out an overall ontology of social life, offering concepts that will grasp both the rich texture of human action and the diverse properties of social institutions. Some of these concepts should be useful sensitizing devices for research purposes, while others help provide an explication of the logic of research into human social activities and cultural products.
>
> (Giddens 1990, pp. 310–11)

Some commentators claim that he writes in an excessively obscure manner:

> the systematic avoidance of examples that might tell what the abstract concepts mean; diagrams whose elements are words and whose spatial organisation has no discernible meaning; citations almost entirely to work of one's own that is at the same level of abstract emptiness.
>
> (Stinchcombe 1990, p. 47)

Giddens sets out three principles as relevant to the overall orientation of research in the social sciences (Giddens 1990). This is methodological rather than theoretical guidance. He calls the principles or precepts *contextual sensitivity*, the *complexity of human intentionality* and the *subtlety of social constraint*. Contextual sensitivity refers to the significance of context in any description or explanation; this includes time and space and the knowledgeability of actors. The complexity of human intentionality refers to notions of intended and unintended consequences, while the third precept, the subtlety of social constraint, deals with the mediation of social constraints through agents' reasons.

Context is of overriding significance, according to Giddens. It includes spatial and temporal dimensions and links causality to the knowledgeability of actors, that is to say that one feature of the context can only be understood to be causally linked to another feature to the extent that human agents know about the link.

Giddens argues that structuration theory acts as a sensitising device. The complex nature of the structures and systems involved in a case, and the nature of the actors' actions, cannot be understood simply in terms of individual motivation or psychology, or as part of some overbearing sociological frame. The relationship between action and structure is a very subtle and complex one. These injunctions are well taken in this project. As a sensitising device structuration theory is valuable. But there is a problem that in insisting that causality exists only within the structure–agency relationship and that structures are the instantiations of human actions, Giddens seems to deny the possibility of structures having a separate existence. They have effect only when instantiated in human action. Giddens's structures are reducible to human action. They are 'traces in the mind'. In insisting that social theory avoids the determinism of functional arguments, Giddens constrains himself to a fundamentally hermeneutic position. But as Layder (1987) argues, this does not allow for an objectivist account of social reality. Such an account must operate with a notion of structure as, at least partly, preconstituted and relatively independent of the social practices it partly conditions.

Structuration theory is increasingly being used to inform research in the area of qualitative information systems as a framework for interpretation (Walsham 1993; Orlikowski 1992). The inability of the theory to deal with relatively autonomous features is a problem when discussing technology and this is taken up in later. Structuration theory is taken seriously in this project. The recursive relationship between agents and structures forms a major part of the argument and Giddens's sensitising precepts are recognised. However, when technology is bought into the argument the partial independence from human agency of some explanatory elements must be recognised. The categories adopted by critical realism provide a helpful orientation.

## Critical realism

At one level, critical realism is a common-sense form of argument. Baert (1996) sees it as equivalent to M. Jourdan's realisation, in Molière's *Le Bourgeois gentilhomme*, that he has been speaking prose all his life. But it has a distinctive methodology which can be quite precise (Keat and Urry 1975; Bhaskar 1979; Baert 1996; Lawson 1997) and which uses many of the insights of structuration theory, although the two developed in parallel, and largely independently. In economics, critical realism has developed in opposition to positivist accounts using closed systems. It thus has attractions for those interested in the importance of institutions since it can potentially provide a theoretical and methodological framework more consistent with the framing assumptions outlined above.

Empirical realism attributes reality only to empirical entities. Scientific realism asserts that the ultimate objects of scientific observation exist for the most part independent of, or at least prior to, their investigation. *Critical* realism has a structure of three domains (empirical realism has only the first two): the actual, the empirical and the non-actual. The actual is the events and states of affairs themselves. The empirical is our experience of these events. The non-actual or deep domain refers to the structures, mechanisms, powers and tendencies which govern events. The non-actual is not necessarily accessible to observation. Furthermore, our experience of events is often out of phase, that is events are often unsynchronised with the mechanisms or powers which govern them; thus it does not follow that causal mechanisms can be reduced to constant event conjunctions. Explanations, it is argued, should refer to structures, mechanisms, powers or tendencies and to open systems, that is systems which do not have a closure condition such as equilibrium in neo-classical economics. Individuals in open systems have freedom to choose, but have mechanisms, powers and tendencies operating upon them. Open systems are those in which different outcomes are possible. There are no closed, law-like powers which determine, for example, the market equilibrium. Thus the possibility of choice in human action is recognised as a key assumption of the methodological framework. Tendencies are transfactual statements about what is going on at a deep level, irrespective of the actual outcome; for example, gravity is acting even when an object is held. A tendency, on this definition, is thus an unconditional statement about a power that is being exercised, whatever events ensue.

These distinctions are helpful. It is an assumption of the analysis undertaken in this project that the world is open. The concept of a tendency is also valuable. A routine, as the term is used here, is the manifestation of a tendency. It is the empirical referent of a tendency. Routines are recurring practices but there is no certainty that they will be adopted. Factors may be operating which preclude their adoption. The tendency to act in a particular routine way will not have disappeared but will have been counteracted by other features.

In critical realism, inductive and deductive law-like statements are rejected in favour of retroduction (or abduction, an idea drawn from, among others, Peirce (Mirowski 1987), one of the founding figures of 'old' institutionalism), in which explanation is made by drawing attention to metaphors or analogies with mechanisms which are familiar, in order to understand new phenomena. The metaphor can never be completely appropriate. It is a way of using existing knowledge to contemplate explanations of phenomena less understood. In this formulation metaphors are essential to the conception and development of scientific theories (Lewis 1999).

Metaphors have been used widely in organisational theory as a way of providing interesting insights into new interpretations, new ways of thinking and of seeing (Morgan 1986). Walsham (1991b) uses them directly, though rather mechanistically, to interpret information systems development. The use of metaphors both by actors and those interpreting their actions is a part of the nature of interpretation in any analysis. However, it is important that metaphors are used carefully and self-consciously. They can suggest implications about relationships and processes which need to be consciously assessed. This is true particularly when human characteristics are used in interpreting organisations or institutions, using metaphors of learning, memory and acting, for example.

People are seen as knowledgeable in critical realism but knowledge tends to be practical and embedded in taken-for-granted activities. People's knowledge is bounded by unacknowledged conditions or unforeseen effects. In the research reported here interpretative conventions are used to uncover meanings. While meanings can be uncovered through interpreting human action, and action then acts back on the meanings themselves, it is not necessary to argue that social action and structure are irretrievably hermeneutic.

The approach adopted in the following chapters argues that the persistence of social structures is dependent on the practices or activities which the structures help to constitute. Structures do not exist independently of the conceptions or definitions made by individuals. And social structures are only *relatively* enduring. People can change them, hence the dynamic nature of social life. But these structures can have an existence which is independent of any individual and the instantiations of actions in which that individual is involved. Rules and the hierarchy of relationships associated with them – the structure in Giddens's terms – interact in a recursive way with human agency. Social rules are drawn upon as generalised procedures of action. They govern, condition, limit and facilitate but cannot be reduced to action (Lawson 1997).

In structuration theory and critical realism there is considerable agreement on the nature of the social world. The primary difference for critical realists is that the objects of science are structured in the sense of being irreducible to events and their patterns. Mechanisms, tendencies and structures are set out as causal factors and are referred to as *intransitive*, acting independently of the process of their identification. They are irreducible to our knowledge of them and in some part endure and act independently of our knowledge of them.

Causal factors can be unobservable – part of the non-actual world. Such an approach is commonplace in the natural sciences where the existence of such factors is deduced from their effects. Analysis of this kind enables us to look for tendencies and structures and to attempt to explain them. The explanation of routines here draws on these categories and adopts the realist ontology: the objects under study in some part endure independently of our knowledge of them. As an orientation or logic with which to analyse social activities, structuration theory can provide helpful insights which warn against simple unidimensional explanations. It has been used helpfully in the information systems literature and in developing a theory of technology but it struggles to address the material nature of technology which a realist interpretation can more readily take into account, and it is to this that I now turn.

## Analysing technology

The research looks at change in organisations to try to understand the way in which routine behaviour affects change and is affected by change.

Technology is a major focus both because it is a key exogenous variable used in orthodox economics to account for change in firms, and because strongly technological-determinist accounts are commonplace in common-sense discourses about change in advanced economies and require investigation. The research endogenises technology by building into the account a theory of technology and the way it is used which is compatible with the underlying theory of change processes.

Some commentaries imply that we are undergoing a new industrial revolution. Does technology drive – or is it merely an enabler? Could the changes we see have occurred (differently) without electronic systems? A fascinating and thoughtful source which reflects this discussion is Zuboff (1988). She shows how the potential of information systems makes visible previously hidden aspects of organisational processes and may have a significant effect on the way individuals learn about their work:

> To fully grasp the way in which a major new technology can change the world . . . it is necessary to consider both the manner in which it creates intrinsically new qualities of experience and the way in which new possibilities are engaged by the often conflicting demands of social, political and economic interests in order to produce a 'choice'. To concentrate only on intrinsic change and the texture of an emergent mentality is to ignore the real weight of history and the diversity of interests that pervade collective behaviour. However, to narrow all discussion of technological change to the play of these interests overlooks the essential power of technology to reorder the rules of the game and thus our experience as players.
>
> (Zuboff 1988, p. 389)

These issues have only relatively recently been seen as an important element in understanding the kinds of electronic information systems which feature in this research. In early work in this area information systems were seen as part of the technical arena. Work was concentrated around issues of design and development, and even when processes of use were discussed they tended to be seen as primarily technical. Lewin (1952), for example, suggested that implementation of information systems could be seen as a three-phase process involving 'unfreezing', 'moving' and 'refreezing'. The process outlined was a simple one and referred essentially to the assessment of needs and problem definitions (unfreezing), the definition of objectives and solutions (moving), and the evaluation of new systems and transfer to the client (refreezing) (Ginsberg 1979). Implementation was seen in terms of organisational change but this was interpreted as getting the results of information system work into the hands, or minds, of managers so that decision making by those managers changed (Schultz and Slevin 1979). The model used was compatible with Chandler's view of change processes, and the implementation of predetermined patterns was the central concern.

Such an approach was challenged as resistance to information systems was experienced in the 1960s and 1970s. The key developments which appeared in response to such challenges initially involved an examination of the complexity of the implementation process itself. Ginsberg (1979) emphasised that the diverse perspectives of project participants should be taken into account. Their judgement of success was seen as a key measure. In particular, Ginsberg argued that management scientists should not be the only source of data in research as they had been hitherto. However, it was not until the 1980s that social rather than technical issues were picked up (Keen 1981; Markus 1983). Keen noted the pluralism of organisational decision making and commented upon the link between information and power. He ended with a plea for more political studies of information systems and argued that case studies are a legitimate and much needed part of IS research. Markus focused on political dimensions. She examined a number of theoretical perspectives which differed in their assumptions about systems and organisations, and concluded that an explanation which considered the complex interaction of the system and the context of its use was to be preferred. Markus clearly brought a socio-political dimension into the mainstream. Her analysis recognised different interests within organisations.

Kling and Iacono (1984) adopted a more complex socio-political approach. They demonstrated that key actors manipulate organisational ideologies in order to increase power and control. Kling and Iacono comment upon understanding the *intentions* of key players and are in no doubt that interests and meanings within an organisation are varied and relevant to information systems. The emphasis on participants and users which developed at the beginning of the 1980s was taken as the central focus in later work. Srinivasan and Davis (1987) argued that process models are themselves inappropriate and are incapable of capturing the implementation issues involved in the latest technologies. Such considerations also allow the analysis to be broadened still further to include notions of equity and evaluation. Lyttinen and Hirschheim (1987), Symons (1990), Joshi (1991) and Walsham (1991a) in different ways all pick up on the broader issues. Symons's concept of multiple perspectives, which is also adopted by Walsham, puts the perspectives of participants, and the way they make sense of their worlds, centre stage but analyses them through an interesting range of multiple theoretical approaches. This builds on the earlier approaches though Symons's theoretical perspectives come from a wider set of traditions and she takes a deeper and more sophisticated stance.

Theoretically and methodologically the approach of analysts has become more complex in terms both of the factors seen to be relevant and the way in which they are studied. The philosophically simple (though frequently technically complex) models which were in use during the first two decades concentrated on the solution of basic, technical computing issues. Simultaneously, as these issues became more tractable, so organisational issues became more compelling. Users were less willing to accept neat technical solutions to somebody else's problems and so attention switched to the human, social and political dimensions.

Thus the analysis of electronic systems within organisations is now seen to require a consideration of the interplay of technical, social and historical factors. Zuboff (1988) sets out the complexity of this kind of approach in the quotation cited above. In her analysis she has invented the term 'informating' to characterise some of the interrelationships. Informating has parallels in automating. While automating took mechanical effort out of production and changed the lives of workers, informating changes the relationship between the worker and the product by creating new qualities of experience and by making visible information which was previously unknown or known only to a limited group.

Alongside this, other reflections on information systems point out the benefits for learning of, for example, computer-assisted methods or flight simulations – codifying what was previously an uncodified skill – or point to the potential of artificial intelligence systems suggesting that learning is instantiated not just in the information systems but in the information technology itself – that we may be able to build intelligent systems (Jones 1994). The nature of agency is questioned here. If the information technology itself, in some senses, embodies learning it becomes necessary to consider the extent to which technology can be considered an agent in its own right.

Orlikowski (1992) sets out the argument for using a structurationist approach in carrying out this kind of analysis. She looks at the dualism of material and social, represented by the human use of technical systems, and shows how the relationship between agency and structure is a recursive one. In this analysis routines can be considered as (incompletely) institutionalised practices formed and changed within a structurationist interpretation. Orlikowski's arguments are fluent and persuasive. Her use of structuration and similar attempts to integrate these theories into established analytical approaches to information systems (Walsham 1993) give a useful model on which to approach the complex technical and social interactions found in information systems. A number of different approaches are used. Walsham is a major and articulate proponent of the use of concepts from structuration theory and for the use of structuration as a meta-theory, taking Giddens's claims for structuration as a sensitising device seriously. He has used concepts from structuration theory in analysing a number of cases and, at the meta-theory level, has used structuration as a perspective 'within which to locate, interpret and illuminate other approaches' (Walsham and Han 1993, p. 81). While this breaks away from technologically deterministic accounts, it implies the existence of a taken-for-granted knowledge which different theoretical positions can help to uncover. It is important, however, in using ideas as illuminatory devices on case material to recognise the theory-laden nature of the research process itself. The epistemological implications of this must be recognised. Giddens argues that epistemology is unimportant and that structuration theory is trying to establish the ontology of social facts, which is fine if the objective is to sensitise. But if these ideas are used in social *research*, knowledge claims are important. We cannot simply apply structuration theory to established facts because the choice of those facts is determined by the theoretical position and values of the researcher.

Orlikowski (1992) effectively recognises the epistemological issues by explicitly making structuration theory the underpinning framework of her analysis. She sets out a duality of technology which parallels Giddens's duality of structure. The duality of technology identifies technology as a product of human action. Technology is physically constructed by actors working in a social context and socially constructed by actors through the meanings they attach to it and by the ways in which they choose to use and deploy it. The structurational model of technology then has four parts. Firstly, technology is the outcome of human action such as design and development; secondly, it facilitates and constrains human action through the provision of interpretative schemes, facilities and norms; thirdly, institutional properties influence human interaction with technology, through, for example, intentions, professional norms and resource availability; fourthly, interaction with technology influences the institutional properties of organisations through reinforcing or transforming structures of signification, power and legitimation. This model thus builds a layer into the conventional structurational approach around the significance of technology. Within the conventional approach technology would be included in the two categories of structure and agency. By specifying the model more precisely, Orlikowski creates insight into the way in which technology operates as a constraining, enabling and legitimating device, and she is able to emphasise the inherently social nature of technology. Furthermore, given the propensity for modern information technologies to break down conventional ideas of space and time, and given Giddens's concern to encompass space and time, there is a potential for rich analysis.

There are difficulties, here, too, however (Jones 1998a). They arise from the material nature of technology as defined by Orlikowski. If technology is a material entity it is more than, or different from, the traces in the mind of Giddens's structures. The material nature of technology, while still open to interpretation, cannot be reduced to the instantiations of agents' actions. Routines, norms or rules which are embedded in the technology, but which are recursively flexible, similarly have a real existence. Orlikowski recognises this difficulty (Orlikowski 1995) and distinguishes between technologies as artefacts and technologies-in-use. Technologies-in-use refer to the patterned interaction of a technological artefact and human action, the technological artefact being the physical nature of the technology. This is a useful distinction. It enables Orlikowski to show that the same physical technology can be used in significantly different ways and that the differences which amount to a different kind of technology (in use) are accounted for by the differences in human systems and agency in the two cases. Once the differences-in-use become embedded in the physical artefact, however (in Orlikowski's example this would involve physical developments in the use of the software, Lotus Notes), then the distinction no longer holds. A different technology-in-use has produced a different artefact. The analytical framework showing recursive relationships between structures and agents' actions remains appropriate but it has to be seen as having an existence separate from the human agents involved.

Technology, in the form of electronic systems, is implicated in many of the changes observed in the firms studied in this project. For all the companies it was a significant feature but was then used and interpreted by them in different ways. The technology cannot be interpreted independently of the people who work with it (Latour 1996a, 1996b). Latour contrasts technical and social determinism and argues that a synthesis is vital: 'The first one took humans as irrelevant; the second tried to circumscribe the non-humans as much as possible. The third follows as far as possible, and in all its consequences, the impossibility of allocating humanity and non-humanity in the first place' (Latour 1996b, p. 299). Different views existed in the companies about the value of the technology, and the companies found themselves pulled in different directions. Disagreements were not usually explicit. They arose as grumbles or comments on how things might be. In some cases the use of modern information technology was so much a part of the way of doing things that no alternative approach was considered. Technology in this sense was not a technical feature but a demand for service or a form of relationship. Technology as artefact and technology-in-use (Orlikowski 1995) were inextricably mixed. The technology did not so much have an impact on organisational routines (and the use of the word 'impact' has an implicit element of technological determinism), rather it was constitutive of them. That is not to say that technology was more important than human behaviour but that an attempt to separate technology from human practices and meanings diminishes the importance of both and fails to capture the inextricable connections between them.

Thus, in the companies analysed here, the appropriate interpretative framework is neither technologically determinist nor socially determinist. The material and social features of the cases interact so that human action is constrained and enabled by the technology but the technology is interpreted and changed by human agency. Jones (1998b), following Pickering (1995), argues that this approach incorporates material agency in the 'double mangle' of human agency and interpretation. That is, social structure is not embedded in technology in an inflexible way, nor is there an imputation that machines can possess intentionality. Human interpretation can be seen, in the cases which follow, as central to the design, implementation and use of information systems. The interpretations depend upon the social and technical interactions which are themselves path-dependent. The character of technological systems is 'as ongoing artefact-in-construction through the situated practice of knowledgeable agents' (Jones 1998b, p. 16).

# 3 Research methods

## Introduction

The theoretical underpinnings of the project demand a research method which is able to investigate deep structures and the practices associated with them. Conventional econometric techniques focus on empirical events, usually attempting to draw explanations from assumed constant event conjunctions. Such techniques are inappropriate in a project of this kind. We are looking for patterns in behaviour, recognising that such patterns may be influenced by a range of factors and different events and that actual outcomes are open even though there may be a regularity in the way in which the firms studied behave.

At the same time the primary research evidence lies in the interpretations of individuals working in the research sites. It is important not to prejudge the events and relationships which are important for them, or to force their interpretations into some predetermined framework.

Qualitative analytical approaches can fulfil these requirements. Such work has been carried out in the social sciences for a considerable period, but it has only relatively recently achieved any prominence outside cultural anthropology, as doubts have arisen over the epistemological claims of the crude use of quantitative methods, and qualitative methods have become accepted (Van Maanen 1979a). Interpretative work, as part of qualitative methodology, has evolved slowly (Altheide and Johnson 1994). By its nature, each interpretative study will be different, but there is now a corpus of practice upon which economists can draw in order to acquire a rich picture of economic institutions. With case studies providing one form of the interpretative method, the detail will be specific and relate to a particular case, but it can be used to build theory (Eisenhardt 1989) and to improve understanding (Van Maanen 1979b); nor is it incompatible with other research methods (Hari Das 1983; Bryman 1989).

The research process is different from that typically carried out in orthodox econometric methodologies. In particular, analysis is not primarily the last phase of the research (Tesch 1990). Analysis and data gathering are carried on in parallel and the research is grounded in its object of study (Glaser and Strauss 1967). As in all intellectual activity, interpretative analysis is both systematic and reflective, but here the researcher needs to be particularly self-conscious about the nature of the research and to reflect on it as it proceeds. The researcher is attempting to make sense of the interpretations of others and must be aware of the influence of himself or herself on the research process. Consciousness of the context of the research and the cognitive models held by those contributing is also vital. Once collected, data – in this case primarily research interviews – are segmented, categorised and compared. The comparisons are used to find conceptual similarities and to refine categories that grow out of the research. Manipulation of the data is relatively eclectic, requiring leaps of imagination, though there are systematic procedures that can guide the analysis to a higher-level synthesis identifying patterns, themes, structures or fundamental practices. One such set of practices could be a system of routines, but it is important in work of this kind not to impose that framework on the data collection. Patterns arise from the data and are interpreted by the researcher alongside other frames of reference.

Fieldwork and its interpretation can explore the meaning of terms and allow us to develop, elaborate and challenge the concepts used in theoretical accounts. Theoretical categories have to be related to the way people conceptualise their own economic behaviour. The underlying question is not so much 'how much does this change?' or 'what is the precise relationship between this and that?' but rather 'what have we here?'

Different fieldwork methods are available, ranging from participant observation to pre-coded questionnaires. The method adopted here falls at the relatively unstructured end of the spectrum. It is predicated on a requirement to make sense of the interpretations of others rather than to test a preconceived explanatory framework. The most appropriate fieldwork method for this project was a series of lightly structured interviews and observations. Studying only a limited number of cases made it possible to develop a rich picture of each case. Semi-structured interviewing allowed the respondents to participate in the development of the research agenda.

The research method itself broadly follows the precepts set out by the proponents of grounded theory (see, for example, Strauss 1987). This approach has developed a distinctive series of procedures of its own but there are no hard and fast rules. It recognises that research data can be of many kinds, including interview transcripts, documentary evidence and balance sheets, for example, and that social phenomena are complex. Importantly, grounded approaches are premised on a search for theory, not simply description. They argue that without grounding in data, theory will simply be speculative.

Grounding refers to the need to refer back continually to the data. Patterns and hypotheses arise from the data. Particular concepts appear to be important but before any theoretical assertions are made the emerging explanations must be tested against the data, time and again. Coding data and comparing them are the primary features of this approach.

The early work in grounded theory occurred during the 1960s (Glaser and Strauss 1967) but Strauss acknowledges connections with the school of American Pragmatism, in particular the work of Peirce, the institutionalist philosopher whose concept of abduction emphasised the role of experience in the early phases of research. The selection of data is thus not purely inductive or deductive but comes

> from experience with this kind of phenomenon before – whether the experience is personal, or derives more 'professionally' from actual exploratory research into the phenomenon or from a previous research program, or from theoretical sensitivity because of the researcher's knowledge of technical literature.
>
> (Strauss 1987, p. 12)

Abduction or retroduction thus forms an important part of the practical, detailed work of research of this kind. The approach of grounded theory thus sits comfortably with the methodological position outlined in Chapter 2. It is one way of attempting to uncover the deeper levels of causation in social phenomena.

A final, important part of ethnographic styles of research is accounting for ourselves (Altheide and Johnson 1994). The remainder of this chapter sets out the way in which the grounded approach was carried out in this project. This is both to exemplify the method for those unfamiliar with it and to account for myself, following Altheide and Johnson's injunctions.

## The selection of research sites

Since the research was concerned with discovering routines, it was necessary to investigate firms which were undergoing change so that any unchanging behaviour could not be explained simply as a static equilibrium or on grounds of compatibility with an unchanging environment. Relatively enduring characteristics had to stand out from the changing world of the firms. If the organisations were not undergoing change, in some senses all behaviour could in principle be considered routine. The same things would tend to be done in much the same way on each day or month or year. But such routine behaviour is not necessarily an important practice within the organisation; it may be easy to change and have little implication for the way the organisation's members view the way they do things. When the environment is changing, routines, if they existed, as the established, significant, sanctioned and recurrent practices within organisations, would stand out as important features. By definition, they would be the things which do not change or change much more slowly than other forms of behaviour. There should be a regularity and predictability about them.

A focus of the project was the implication of changing technologies for firms. A range of forms of engagement with technology was therefore required, and because the research was concentrating on change, a longitudinal approach was appropriate in order to observe change and to seek out reasons for the changes and the enduring characteristics of the companies. Each firm would be studied over a period of about two and a half years in order to observe the changes the firms underwent and to analyse the changing relationships and structures within them.

In planning the research therefore I decided to concentrate only on firms in one geographical area: Cambridge and the small towns and villages which surround it. This meant that any differences between the firms were unlikely to derive from geographical variations. Furthermore, because the Cambridge area is noted for the development of high technology it was likely to be relatively straightforward to find firms engaged in high-tech work. The particular area of high technology chosen, because of the growth of interest in it and its apparent potential for change, was electronic technologies linked to the dissemination and diffusion of information.

The research sites were selected after a good deal of searching. From the beginning, selection was based on a limited number of criteria. The firms chosen had to offer potentially interesting forms of behaviour related to high technology. There should be differences between the firms but, if possible, similar characteristics too, so that contrasts and comparisons could be drawn.

The focus was on small or medium-sized enterprises employing fifty to 100 people so as to make it possible to get a sense of the organisation and to speak to employees throughout each organisation. It would also thus be possible to compare companies without interpretations of their behaviour being swamped by a big-company effect based on experiences remote from the research sites though the latter objective failed in one case.

The four firms which participated in the research were Digital Equipment Company Ltd, National Extension College Trust Ltd, Unipalm Group PLC and Chadwyck-Healey Ltd.

After getting in touch with the senior person at each company by letter, I arranged a visit to discuss the particular interests I had. Both parties could then decide whether collaboration over a two- or three-year period was potentially interesting. In each case I gave a brief presentation of what I intended and a short written summary. I asked the senior person for background on the organisation and for any reports or other published papers which might be available, and took extensive notes.

The purpose of these initial establishing interviews was to obtain background information about the companies to make sure they were suitable research sites and to build confidence within the company that I would not interfere in their work and that discussing policies and taking time out might be helpful to them, providing a sounding board or an opportunity to clarify thoughts.

In addition to setting out the broad objectives of the project, I also set out the way in which it would proceed in future: I should visit the company regularly and interview a group of staff. The group would include the senior person and an employee on a tier below but, if possible, not working directly for him or her. Interviews would also be carried out with employees in tiers further below, each reporting to a different person from the one interviewed in the tier above. Thus it would be possible to get a sense of the breadth and depth of the organisation – a 'diagonal slice'. I asked the senior person to recommend suitable colleagues and in all cases they came up with a list of people willing to join in.

The participants were thus selected by the senior person. Potentially, the senior person could have selected a sample of people who had a particular view of the company. I had no easy way of selecting on any other criterion but, in practice, the individuals selected were all people who were happy to express their own views and did not take a company line or act as company apologists. Asking for a diagonal slice was one way of making the selection of respondents relatively open since the senior person was unlikely to have detailed contact across that range. I also asked that a mix of relatively recent employees and people of longer vintage should be selected if at all possible; in the main this was achieved, thus giving potentially a range of views and experiences and limiting the opportunity to 'fix' the sample. The sample of respondents in all cases did not appear to have been chosen for any reasons beyond those specified, though clearly such a selection process almost certainly omits any notorious or maverick views.

## The interviews

Once the initial relationship with each company had been established, the research followed a pattern broadly in line with that recommended by grounded theory. Strauss (1987) lists the main elements and research phases he regards as appropriate for grounded theory. His lists tend to be mechanistic, lacking the flexibility which is required in practice, and he sets out an idealised version of what can be a complicated process requiring subtle judgements and an ability to think on one's feet. But as a framework his recommendations are helpful.

There were three rounds of interviews after the first establishing meetings. The first set of interviews was aimed primarily at developing categories. Analysis of the transcripts showed that a number of categories were emerging. The second round continued the same broad pattern but identified changes where these had occurred, and by building on the developing relationships with the interviewees sought to get behind any facade which had been erected. The third round again continued that pattern but also checked the emerging categories with participants.

Interviews in the first round were informal and lightly structured. The intention was to find out what respondents believed was going on, what was driving change and why. Specific questions were asked about information technology and respondents were asked to reflect on the next six months. Interviews were usually carried out in a meeting room rather than at the respondents' work location though the senior persons at NEC and at Chadwyck-Healey both preferred, and were interviewed in, their own offices. Interviews with Digital's homeworkers were carried out in the foyer of a nearby hotel or in a 'touch-down office'.

None of the interviews was difficult in personal terms. Each respondent had been asked to take part and had agreed. Each was interested to find out what the objectives of the research were. Each interview began with a very general explanation about looking at change in organisations, not a list of specific objectives which would have pre-empted the agenda. All interviewees agreed to be recorded and for their names to be used in any analysis. (Occasionally, in subsequent interviews, some participants asked for items to be discussed off the record.) The intention was to get interviewees to talk about themselves and the work they did and through this to reveal their understandings and definitions. Respondents were happy to do this and interviews frequently went on beyond the time which had been notionally set aside. I did not offer opinions or critical comment. The interview style was as open as possible with comments put back to the interviewees, such as 'So you did that', in order to encourage further exploration of comments. It was important that the respondents were able to determine the issues they believed to be important.

Before the first round of interviews only a limited amount of desk research was carried out on each company. I wanted to have a sense of the companies so that I could come across as knowledgeable, with an intelligent interest in their affairs, but I did not want to prejudge the issues which would come out of the interviews. Consequently most of the background reading took place after the first interviews and was partly informed by them.

Once the interviews at a company had been completed, each respondent was sent a summary of the notes on the interview, along with a letter of thanks and a request for comments or corrections. These notes did not include asides about possible connections between firms or individuals or possible theoretical explanations, nor did they include any sense of the attempts which were made by respondents to position themselves in the story. The notes were simply a record of the comments which had been made by the respondents. Whenever possible, the notes and thanks were sent by e-mail. E-mail felt less formal, made relationships more easy-going and made use of one of the focal points of the research, which seemed an appropriate thing to do.

The interviews were then transcribed and analysed, and coded using computer database software suitable for non-numerical qualitative data. The coding was modified with use, and the activity of coding and segmenting the interviews acted as a valuable tool for focusing ideas and driving the work.

The second set of interviews took place roughly six months later. Questions were based around what had happened since the last visit, and included some reminders of the predictions made by participants at that visit. Participants were asked why change had taken place, what things had not happened which might have happened, how their jobs had changed and why, where power and influence lay, how good communications were and what impact electronic communications had, what might happen in future and why.

Relationships with respondents felt very easy. The interviews were relaxed and I was treated like an old (and I believed trusted) friend. Respondents were less guarded. Interviews took place in similar locations to those used previously. Less care was taken (by respondents) in determining the location, reflecting their more relaxed attitude and the reduction in their desire to impress.

On the basis of a thorough, detailed reading of all the interview transcripts and attempts to find patterns in the behaviours described, a story was prepared for each company and secondary sources such as company accounts, web-sites, newspaper articles and financial summaries were consulted. A deeper analysis of the database was carried out and further consideration of the research interviews was undertaken, tacking backwards and forwards between the data and the developing story. This was an iterative process.

In writing stories I was acknowledging the impossibility of an entirely objective or scientific discourse. The economic and other theoretical categories were also being placed in the narrative. Strassmann and Polanyi (1995) argue that reconceptualising economic practice as story telling is important. Here I attempted to be self-conscious of the situated position of the knowledge I was creating.

The stories made use of the participants' own words. I selected which to use on the basis of my interpretations of the participants' accounts. However, in view of Hall's claim that language is a privileged medium (Hall 1997), it was important to allow the participants to make their own statements as much as possible, and to be open about the data which were being drawn upon.

The stories were developed one by one. As each story was completed a third round of interviews was carried out. The initial questions were similar to those in the second round but the participants were asked to make comments upon my initial attempts at analysis. Did the ideas or metaphors seem right to them?

Relationships were very comfortable. I felt I was being given 'honest' answers, while recognising that there was some positioning going on. I was able to argue and challenge participants, sometimes quite forcefully. The more forthright nature of the discussion and the reflections of the interviewees on the categories used enabled me to go back to the emerging explanations and to fill them out with further evidence. Strauss would call this saturating the categories, by which he means 'the core category must be proven over and over again by its prevalent relationship to other categories' (Strauss 1987, p. 35).

The interviews were transcribed, segmented and coded once again, and further analysis using the computer database was carried out. The coding used changed a little after the first round of interviews but the pattern set there remained the basic framework. Some codes were merged with others and some were dropped. The interpretation of some codes changed and the relationship between the different categories shifted as it became clear that some codes were subordinate to others and others largely independent.

After completing the final rounds of interviews the company stories were revisited. They were modified in the light of the data available and then taken apart to emphasise patterns which related to the development and reproduction of routines. At this stage the chronological account of what happened in each company was less important than establishing the nature of the routines there and comparing and contrasting the experiences and practices of the different companies. These accounts were then developed to trace different features of routines and their reproduction through the four companies. Subsequent chapters pick this up. Each is situated in a particular company and carries a specific responsibility for the development of the argument about the nature of routines in companies like these.

## Reflections on method

The interpretative methodology required me to be self-conscious about my position in the collection and interpretation of data, which, as set out above, were interconnected, the analysis emerging from continual comparisons of data and possible explanations. It also required me to stand back from the accounts given by the respondents and to interpret them according to my sense of the different messages they were giving about themselves and the organisation. The diagonal slice of respondents in each company gave me opportunities to check accounts from different perspectives and to compare them with informal conversations and published documents.

The interviews were all carried out using an approach designed to elicit information. In briefing the respondents, I explained that I wanted them to tell me what they believed was going on, and that I had no pre-formed agenda. Some junior members of staff were surprised that I was interested in their ideas; in fact, in trying to unpack their taken-for-granted assumptions, I often found what appeared to be the key to underlying structures and cultural meanings. Junior staff tended to reveal their sense of the organisation more readily through their statements about what went on. One more senior respondent commented that he never prepared anything for our meetings and that he simply told me what he felt to be true. He said this half apologetically, but since it was this sense of the taken-for-granted, sanctioned and established practice that I was trying to discover, such comments were reassuring.

My sense of my own position in the data collection was clear to me, and I felt, increasingly as the research proceeded, that the respondents were comfortable with it. I did not claim to be knowledgeable about their business and never passed judgements on the activities I observed or matters discussed. Even in the third round, where I was testing out ideas, it was to seek information. Questions were then along the lines: 'Does this concept match your understanding of that issue?' or 'Can you really justify that sense of the way this company operates? Can you give me examples'.

On some occasions I perceived a wish by some respondents to be seen in a particular light. One respondent, for example, affected the position of a cynical commentator, another appeared to me to be positioning himself as a go-getting, successful and dynamic innovator. In general, however, positioning was subtle and implicit. Some staff simply enjoyed talking about their work. Their positioning was then, in many cases, no more than that in normal social intercourse between acquaintances who trust each other and do not have a big investment in the relationship.

This should not be taken to imply that all interviews and interviewing relationships were problem free. I found one interviewee a very warm character and it was easy to be seduced by him and to use his interpretation of events as the true one. In a different company I was somewhat overawed by the senior person and found it hard to avoid being deferential. A third participant was extraordinarily dynamic, and this too was very seductive. With hindsight, the extent to which I allowed the participants to allocate me the role of friend, supplicant and excited onlooker enabled me to gain their confidence and increase their trust. In examining their statements, however, it was essential to withdraw from those roles.

It is important also, therefore, in research of this kind to be self-conscious about the role of the researcher in the interpretation of the data. I was trying to identify a rich and complex social ontology. Richness and complexity were not difficult to find but in imposing patterns on the data I had to avoid reinterpreting my observations to match the theoretical categories that appealed to me simply because I found them intellectually satisfying. This is not an easy task. There are no lumps of social material waiting to be picked up and examined. The stuff of the research rests in interpretations made by the researcher of interpretations given by the respondents in fleeting moments. In many cases these interpretations did not seem significant until several months after the interviews had taken place (when the examination of the recordings and transcripts of the interviews was undertaken).

The continual, relentless rewriting of the stories of each company was central to the development of explanatory patterns. To obviate the risk that the patterns might become theoretical figments of imagination, the stories had to be regularly checked back against the data. Going backwards and forwards between the data and the emerging explanations is an important part of the approach. The explanatory framework is thus continually regrounded. The refining of categories took away some of the complexity and subtlety of the fieldwork observations but slowly the patterns set out in the subsequent chapters took shape. A number of categories and explanatory frameworks changed significantly during this process. Critical realist categories, in particular the idea of a tendency, slowly emerged as a helpful orientation though the distinctions in many critical realist accounts

(Lawson 1997) are sharper than those that can be made by observation. In particular, the problem for social theorists of taking into account the material nature of technology kept reappearing. In the fieldwork it seemed clear that technology was implicated in human action beyond an explanation concerned solely with the human interpretation of technological artefacts and that an interpretative account, on its own, could not be adequate.

# 4 Digital: the evolution of new routines

## Introduction

In the mid-1990s, the East of England office of the Digital Equipment Company was undergoing considerable change. In coping with that change it fundamentally reorganised its way of working. From operating in a conventional office block, it moved to a structure based upon staff working from home. The senior manager in the East of England at the time of the change described it like this: 'this new way of working was a great imposition on people's lives, people who'd been working with one modus operandi for thirty or forty years were suddenly being asked to do, in their life, something completely different. But to work in a way that was going to impact, you know, their whole life, you know. And it wasn't just your thirty-seven and a half hours a week. It was going to impose seven days, twenty-four hours, either by virtue of a change in use of part of the house or whatever.'

How was the company able to achieve that significant shift and to make 'a great imposition on people's lives . . . being asked to do . . . something completely different'? Staff were required to adopt different practices. Getting out of bed in the morning and going to work felt very different in the new circumstances. We shall see how the company evolved its existing routines to incorporate very different ways of working. Some of this was deliberately managed, some evolved through the interplay of staff, organisational structures and technology. I shall examine how the 'new way of working' was incorporated into the company and how *meanings* of what it meant to work at Digital were used to effect the change.

The new working practices which will be examined here are at an operational level. They are concerned with the day-to-day work of staff and the maintenance of the organisational structure, not with the company's place in the wider environment. The company called its new style 'flexible working'. This has a specific meaning at Digital which will be defined later in this section. The chapter will explore whether these shifts amount to a new operational routine. People's lives clearly changed. The micro-routines involving, for example, sitting in a particular office and responding to colleagues' requests in a particular way, altered significantly. But to what extent can the broad set of operational practices be regarded as new? This is the primary focus of the chapter. At a more general level, therefore, it investigates how operational routines are created and reproduced. The Digital case establishes that routines exist and that they have significance for companies. Once that is established here, subsequent chapters take the concept further to investigate its value at strategic levels and in incorporating change.

At the same time as the company was undergoing significant organisational changes, it was also reviewing its product strategy. A surprising contrast came out of that review. The company had been selling 'flexible working' as a consultancy product since the end of the 1980s but in its review of its product portfolio the decision was made to drop flexible working. This chapter will consider that contrast. Analysing the apparent contradiction whereby the company adopted a particular set of working practices – a new routine – while simultaneously dropping the equivalent consultancy product from its product portfolio helps to clarify the different factors involved in the creation and maintenance of routine behaviour. It is not being argued here that dropping the product was a routine action but an analysis of the factors underlying that decision helps us to understand routine behaviour more clearly.

## Background to the company

Digital Equipment Corporation (DEC) began its life in 1957 as a manu-facturer of computer equipment. In the early 1990s both the market it inhabited and the technology it used and produced were changing rapidly. Some of the changes arose partly from its own behaviour as an innovative company.

DEC was founded by Ken Olsen, a technologist and a Quaker, in Maynard, Massachusetts. The company was a spin-off from the Massachusetts Institute of Technology, and Olsen established it on the same kinds of principles as those used by other well-known Quaker family businesses such as Cadbury and Lever, involving a caring, if paternalistic, interest in employees. It was founded upon advanced and innovative technology. For example, it launched the world's first small interactive computer in 1960 and in the late 1990s was still setting world standards with its Alpha system. During the 1960s and 1970s computing was technical, expensive and the province of experts. Digital was very successful during this period and grew massively. Its VAX range launched in 1977 set an industry standard. By 1983 the company employed over 70,000 people world-wide. This grew to a peak of over 125,000 in 1989.

In 1991, however, the company posted its first ever loss, of $617 million. This worsened rapidly. In 1992 the company lost $2,796 million and losses continued until the year ending 1 July 1995 when profits of $121 million were earned. By 1995 the total number of employees had declined to 61,700.

The UK subsidiary is the Digital Equipment *Company* Ltd. It was formed in 1964, and in 1995 employed just over 4,000 people. Together with the

Irish subsidiary (employing only 450 staff), it accounted for nearly 10 per cent of the company's world-wide revenues in 1995 (Digital 1995). The UK company was thus a sizeable part of the overall DEC operation.

In responding to its declining profitability at the beginning of the 1990s, the parent company took a number of steps. A new President and Chief Executive Officer, Robert Palmer, who came from a manufacturing background in IBM, was appointed in October 1992. In 1993 the company was reorganised into business units and there was a realisation that it was over-staffed by industry standards. According to the 1993 Annual Report, Digital was transforming itself into 'a leaner, more responsive, more competitive corporation with a new organization, new technology and – most important – a new focus on the customer'. Furthermore, 'This focus wasn't just imposed by management; it bubbled up through the entire company' (Digital 1993, p. 4). It is easy to be cynical about such rhetoric, but discussions with Digital staff confirm that individuals were given considerable freedom. It was possible for changes to be initiated by staff and for such initiatives to make a difference to work patterns and products offered by the company.

The financial position of the UK company is shown in Table 4.1. Turnover declined from £875 million to £732 million in the five years following the first trading losses for the parent corporation. Losses were high and the overall picture was gloomy. Employee numbers had fallen by more than 40 per cent over the same period. The fieldwork was carried out during the period in which the company was beginning to move back into profit but the staff interviewed were very cautious about drawing any positive conclusions from the improving figures.

Fieldwork was carried out in Newmarket, one of Digital's UK offices. In 1992, the Newmarket office was one of eighteen Digital locations within the UK, and was responsible for company business in the East of England. It carried out conventional sales and service functions and also had a staff of business consultants. The consultants were largely autonomous. They generated their own business as well as picking up leads which came into the company. Consultants were not obliged to provide solutions which used Digital equipment for their customers. Activities were agreed broadly with managers but the primary objective was to carry out work which generated revenue for the company. The fieldwork concentrated on the work of the consultants. The more senior consultants also had management responsibilities. The precise combination of different kinds of work was complex but, in simple terms, a manager had an obligation to generate revenue as well as to manage a group of staff. Thus a consultancy manager was required to carry out consultancy activities as well as to manage the work of colleagues. The overlapping responsibilities of managers for staff and for a technical specialism created complexity.

When the fieldwork started in July 1994, the senior manager in Newmarket was responsible for the whole of the Newmarket operation and was manager for 'flexible working practices' throughout the UK. Consultancy in flexible

Table 4.1 Digital Equipment Co. Ltd, company financial profile during the period of the fieldwork

| | 06/97 | 06/96 | 06/95 | 06/94 | 06/93 | 06/92 | 06/91 | 7-year average |
|---|---|---|---|---|---|---|---|---|
| Turnover | 721,678 | 666,725 | 731,513 | 800,944 | 809,449 | 799,543 | 875,414 | 772,180 |
| Profit before tax | −19,615 | −19,004 | −35,212 | −62,235 | −97,427 | −29,113 | −7,192 | −38,542 |
| Net tangible assets | 166,597 | 152,905 | 188,499 | 226,369 | 296,169 | 349,258 | 374,802 | 250,657 |
| Shareholder funds | 102,225 | 85,520 | 121,840 | 155,885 | 219,143 | 316,946 | 340,371 | 191,704 |
| Profit margin (%) | −2.72 | −2.85 | −4.81 | −7.77 | −12.04 | −3.64 | −0.82 | −4.95 |
| % return on shareholder funds | −19.19 | −22.22 | −28.90 | −39.92 | −44.46 | −9.19 | −2.11 | −23.72 |
| % return on capital employed | −11.77 | −12.43 | −18.68 | −27.49 | −32.90 | −8.34 | −1.92 | −16.21 |
| Number of employees | 3,508 | 3,184 | 4,031 | 4,977 | 5,720 | 6,166 | 6,833 | 4,917 |

Source: FAME Financial Analysis Made Easy Database.

Note
All figures are given in £000 except where stated; columns refer to the financial year ending in the month indicated.

working was relatively labour intensive and required little hardware. Each project was relatively small and therefore, in absolute terms, generated relatively little revenue. Studying the establishment and continuing implementation of flexible working practices, and how these practices became routinised, formed the core of the fieldwork at DEC. 'Flexible working', often referred to as 'teleworking', has a specific meaning there. It does not refer to part-time insecure work but to flexibility in the location in which work is carried out. Responsible for the product portfolio relating to flexible working practices throughout the UK, the senior manager had developed a range of customer-focused consultancy skills. This is sometimes called the soft end of consultancy. The senior manager was a skilled computer technician but his interests had moved to the human–computer interface rather than the hardware.

At the end of 1992, because of the difficulties facing the company, Digital's continued existence in the East of England was being questioned. The senior manager, C, along with other colleagues, proposed that the East of England centre should adopt teleworking practices and agreed that their substantial office block, the lease on which had come up for renewal, would be closed down. This proposal was championed by C who had the consultancy skill, the network of contacts within the company and the personal management skills with his colleagues to persuade DEC to turn the East of England operation into a telecentre, the first telecentre to be opened by Digital and at that time an idea peculiar to Newmarket.

The telecentre opened in 1994, just before the fieldwork began, in new premises above the company's warehouse. In addition to a small co-ordinating staff there were a number of 'touch-down offices' in which teleworkers, who numbered around eighty, could reserve desks for their visits. Staff were provided with advanced communications links in their homes and a structure of group meetings and informal contacts was established to co-ordinate and manage the operation successfully. The move to teleworking was justified primarily in financial terms. Annual accommodation costs for Newmarket dropped from $1 million to $110,000 (interview with the senior manager, January 1995).

When the fieldwork started in 1994 there was a sense of apprehension but some excitement and relief about the new arrangements. Changes were needed to some of the physical and staffing arrangements but within the limits of the building these were in hand. The building itself offered an enormous contrast to its predecessor – several rooms above a warehouse on a small industrial estate in comparison with a large office block near the centre of the town. A number of staff had been working from their homes for some time and the formal institution of the telecentre greatly improved their circumstances. B, a business consultant, described her move to C's area of responsibility in July 1994 as 'exceptionally positive'. However, the financial circumstances of the company were insecure and it was shedding labour as shown in Table 4.1. There was already some uncertainty about the development of flexible working practices as a product, although this was not apparent to me as an observer for some time, and one of the key individuals in this area had taken a severance package to set up on his own.

Around nine months later, the company was just about to open its sixth telecentre based on its success at Newmarket when it finally decided to drop flexible working from its product portfolio. Business consultancy continued, but flexible working was no longer part of it. The new procedures for teleworking were in place and new routines were developing. Because consultancy on flexible working practices was no longer a company product, C, the senior manager, was made redundant in April 1995. It was widely acknowledged that he had been an insightful, supportive and productive manager. The staff who worked with him had great admiration and affection for him and saw his leaving as significantly diminishing in their ability to carry on.

C made himself available for interview over the following eighteen months and continued to comment on his (now outsider's) view of changes at Digital, as well as the potential for flexible working as a product. He set up a consultancy, which he then merged with a bigger group, and continued to offer flexible working practices as part of his own consultancy business.

Teleworking thus became established as an important form of organisation at Digital but, as a consultancy product, flexible working practices were abandoned. Digital had pioneered flexible working as a business solution and had offered it as a part of its consultancy services for over seven years. Simultaneously with its move into teleworking for its own staff, however, the company concentrated its product strategy on hardware.

The rest of the chapter will trace out the development of flexible working as an operational routine and will analyse why the internal process became incorporated while the consultancy product has been abandoned.

(In June 1998 Digital was taken over by Compaq Computer Corporation. At that time Compaq was a Fortune Global 100 company and the second largest computer company in the world. Digital products were absorbed into the Compaq range and the Digital name disappeared.)

## Staff directly participating in the research at Digital

Five staff from the Newmarket centre participated in the research:

C: the senior manager for the location and responsible for flexible working as a product throughout the UK;

M: a highly qualified technical expert with many years experience, who was organisationally responsible to the senior manager for his day-to-day work and managed teams of consultants within his own specialisms;

B: an experienced business consultant, responsible to the senior manager but with no staff management function of her own;

S:   based at the telecentre and responsible for co-ordinating the work of a group of teleworkers; she reported to the senior manager and provided personal assistant support for him, and was made redundant at the same time as him;

R:   who carried out similar functions to S but reported to the technical expert, M.

In practice the responsibilities of the staff changed during the course of the fieldwork and the descriptions above are approximations to give a sense of the relationships and relative positions involved. C, M and B all worked from their homes; in B's case, for example, this was 40 miles from the telecentre. S and R were based at the telecentre.

The departure of the senior manager, C, and telecentre support worker, S, had an impact on the morale and working practices of those remaining which clearly influenced the responses they gave to my questions and the sense they had of the company's culture. It caused disillusionment for B, the business consultant, who found herself without clear management lines; a huge sense of loss for M, the technical expert, who put in enormous effort to maintain the staff support practices which existed before C left; and a feeling of disbelief from the telecentre support worker, R.

## The creation or evolution of a new routine?

Between 1992 and 1996, flexible working or teleworking became established as a standard company practice at Digital. With hindsight it appears to be a rational, clear-sighted choice which saved the company money and enabled staff, who were in any case mobile, to work in ways better suited to their personal circumstances. The staff concerned were sales and support staff plus consultants. Such staff are mobile in the sense that a good part of their work inevitably takes place away from their office base.

But this is too simple an explanation. Teleworking as a concept had to be argued through. There were other more conventional alternatives, such as moving functions to fewer locations, thereby reducing staff numbers and gaining economies by concentrating work in fewer buildings. The financial circumstances presented an opportunity for new working practices to be tried but a simple correlation between financial stringency and new forms of working does not help us to understand why this particular form of working was developed. As set out in Chapter 2, routines are seen here as established, significant, sanctioned and recurrent practices within organisations. How and why did teleworking, rather than some other set of practices, become established, significant and sanctioned at Digital?

Flexible working practices had formed part of the consultancy portfolio for some years and a good deal was known about them before the Newmarket location moved to teleworking. But it was not straightforward to transform a major office location into a telecentre and there were significant implications for the wider company. Teleworking is a pattern of work in which the employee works from home or some other remote location using computer networking and carefully co-ordinated support systems. It makes use of complex computer-driven communication links which are absolutely central to its operation, but the key features of successful teleworking relate at least as much to the organisational procedures and networking facilities (in human terms rather than computer terms) which are developed alongside the communication hardware and software, as they do to the computer power itself.

The senior manager quoted at the beginning of the chapter put it like this: 'this new way of working was a great imposition on people's lives, people who'd been working with one modus operandi for thirty or forty years were suddenly being asked to do, in their life, something completely different. But to work in a way that was going to impact, you know, their whole life, you know. And it wasn't just your thirty-seven and a half hours a week. It was going to impose seven days, twenty-four hours either by virtue of a change in use of part of the house or whatever.'

Then he continued: 'there's absolutely no way you can move forward unless the infrastructure is in place within the organisation, and that isn't a superficial sort of blob that's stuck on the end of your organisation, it's the infrastructure – however you interpret it, it has to be endemic, it has to be seen as a thread throughout your organisation. So in other words the telecentre could be seen as a bespoke piece, a discrete piece rather hanging on the end of Digital, but for me it was endemic because policies, procedures, all the links and threads were right back through into Digital very deeply. It wasn't just a front to the flexible workers. And the development from that onwards would obviously be four and five and six telecentres so it really is endemic, wherever those people went in the UK, as flexible workers within Digital they were recognised and the infrastructure supported them for their telephony and data, voice coms, whatever help they needed. So they could talk to an HR [human resources] manager in Warrington and that HR manager – if they're working there for the day – would understand fully the issues that they'd got and whatever. So that's the next piece.

'And the third for me is the technology. A lot of companies today seem to think that they can move into virtual working by virtue of the technology's here now, isn't it, the internet now makes it possible, doesn't it? Well no, it doesn't. For me I still advocate the technology as an enabler, a simple enabler and it's getting easier and easier to use.'

The senior manager is arguing, then, that teleworking is a major transformation for the individual teleworker. Conventional domestic space becomes work space. Practices which have been followed for decades are changed and the company imposes itself on the worker twenty-four hours a day. From the company perspective, the process has to be fully integrated in all its locations. His use of the term 'endemic' is unusual but appropriate. I interpret his statement to mean that the routines associated with teleworking have to be incorporated into all the company's locations in such a way that they become natural and taken for granted. Teleworkers must be seen as a conventional part of the organisation if they are to be able to carry out their functions satisfactorily and, correspondingly, if the company is to get the best it can from them. Finally, and much more low key, the technology is an enabler. It does not drive the process but enables it to take place.

Such comments are unexpected from a manager whose life has been spent getting to grips with sophisticated technology. His claims were borne out by other staff, however. For example, the business consultant B's early positive attitude derived from the personal and managerial support she was given. For her the technology was taken for granted but not central.

Establishing a telecentre, therefore, required a change in routine behaviour. That is to say, the practices associated with work, which had been followed for years, had to change both for the individual and the company. The-way-we-do-things-around-here was apparently transformed. The culture and structure of the company had to be amenable (or be made amenable) to such changes, as did staff attitudes. But is this the *creation* of a new set of working practices or the *reproduction* and *evolution* of an existing routine? On the face of it there have been major changes. I shall go on to show, however, that the new working practices recognisably emerged from pre-existing routines, though they were themselves major developments. The firm did not reinvent itself. It built on existing, taken-for-granted ways of working.

The need to reassure staff was recognised explicitly from the beginning, according to the senior manager. A great deal of effort was put into persuading and inspiring Newmarket staff, and into developing carefully integrated human and technical systems to make the concept of a telecentre come to fruition. Those staff involved in setting up the new procedures spoke positively about the efforts taken to smooth the changes. Others less centrally involved, such as B, were less aware of the detail, though as already observed, she found the new arrangements 'exceptionally positive'.

Different patterns of working had to be developed giving support to the view of the changes as a new routine. Relatively formal procedures were eventually laid down in February 1994 in a 'Flexible working handbook'. It is interesting to note that these things were not written down until centres other than the East of England were considering the adoption of flexible working practices. Much of the early work was carried out through small groups in the East of England centre working together and devising appropriate methods. C described the initial changes like this: 'But I guess, yes, it was a passion of one person and then within a few weeks a growing band of people, my managers, key consultants who basically started working very proactively and then an encouragement of everybody else really. Obviously when you announce something as severe as "Your building's going to be closed. Within six months you'll have nowhere to work", it's as good as redundancy anyway for most people. This is the end. People find it very difficult to accept it. Let's go and try this, let's try that. You get the fatalists, you get the cynics. Usual people in any organisation. But I think after about two months of open forums, of going through flexible working with people, the clinics that we set up. Sort of every week we had a clinic at lunch-time. And people were able to come in and sit down and discuss it.'

Although the company had operated with a small number of flexible workers for some time, and then still offered flexible working consultancy as a product, moving a whole office to such practices was new and resisted. My interpretation is that the threat of office closure, based in the difficult financial situation of the company, focused minds, so that reluctant people were willing to give it a chance; but the massive change in routine behaviour implied by these proposals required enormous effort and persuasion. In a later conversation, M, the technical expert, was particularly aware of the subtlety of this process and recognised the difficulty of sorting out the problems which the senior manager's redundancy implied. He saw his own function as carrying on the process which C had set up, maintaining, as he saw it, the founding values of the company. Thus while significant changes were undertaken, they were seen as emerging from the founding values of the company.

### Factors implicated in changing working practices

A number of different factors were implicated in the changes taking place. The changes emerged from existing values but the move to teleworking needed a champion who could operate as a strategic agent. The senior manager filled this role.

Secondly, the skills of the strategic agents had to be right. C's attributes were well developed for such circumstances. All the interviewees were clear that he was trusted by his staff. He had worked flexibly himself and developed a consultancy portfolio in this area. He had a good network of contacts in the company. Other staff in the East of England, such as M, who themselves had the trust of the company and were also senior in their own sectors, supported C's proposals. Their support was seen as important by C and was recognised as an important factor by the staff concerned. Thus a mix of technical, sales and consultancy personnel supported the broad principles and operated the detailed procedures. Strategic agency was not the province of one person. Other staff participated and contributed to the overall direction of change.

Thirdly, the need for security and a clear sense of identity within the staff group concerned, as well as within the wider company, was recognised implicitly. The systems which the East of England team developed used the computing power necessary – and relatively easily available – and mimicked the office environment which the staff were used to. Virtual Private Networks (VPN) were used, for example. VPN are telephone systems which enable individuals' home lines to be connected to the network of the company. They then act exactly like conventional company extensions. Initially VPN were not available. Individuals had to phone in to Digital locations in the same way as any other 'outside' person. C claimed that the move to VPN made staff feel a fuller part of the company and meant that people from elsewhere in the company, or from outside, could not distinguish flexible workers from conventional workers through their telephone systems. The technology thus enabled flexible working to be better established. It did not drive the change but made the continuity of it more probable. E-mail and conferencing systems were also set up. Interestingly, relatively little use was made of computer conferencing. It had originally been conceived as a potentially important means of keeping in touch but it did not take off. This is perhaps because it was not part of normal office procedures and did not offer a great deal more than e-mail. Staff had not learned the routines relevant to electronic conferencing prior to the move to teleworking. At more mundane levels, the need to adopt conventional practices was made explicit. The 'Flexible working handbook' (Digital 1994) includes, in its tips for working from home, advice to 'replace the ritual of going to the company office with another ritual such as scanning the news headlines' and not to 'work on a bed or sofa' or 'allow household to expect you to perform domestic tasks during your work hours' or 'forget to stop working'. The senior manager was very formal in his own approach: 'I think, because they've all heard of flexible working or tele-working over there. They ask you what it's like. "Do you feel isolated?" "Does your manager support you?" All the usual sort of questions. "Do you get lots of free time?" They envisage you being at home with jeans and sandals on and it isn't like that. I put my suit on nine o'clock and I work at my desk with my jacket on the back of the chair. It just happens to be the way I like working. No different for me really.'

Other activities were modified as the system settled in. A much higher number of telephone calls were fielded at the telecentre by the co-ordinating staff than had been envisaged. Extra telephone–reception staff were employed. When staff met at the telecentre, formal meetings were found to be better attended and much more productive than before. This was something that all participating staff volunteered early in our conversations about the effects of teleworking. The meetings had an important relevance for the staff.

Informally, the lack of social facilities at the telecentre meant that some effort was made to devise social events. The coffee bar at the local Tesco store (just across the road) became the common room. Staff readily divulged these little details in their asides about what it felt like to work in the new circumstances. They acknowledged perceived needs to socialise and that C's awareness of this meant that managers were making some effort to facilitate such activities. They were not occurring entirely by chance.

Thus informal practices, such as the use of Tesco, and formal practices, as advised in the handbook and facilitated by the technology, were developed. They formed the broad set of working practices which constituted the operational routine for teleworkers at Digital. In general the routine created an environment for flexible workers which matched that of the conventional workplace as closely as possible. The particular pattern appeared not to have been adopted deliberately, but staff found it comfortable and it had emerged and had been built up as the senior manager in particular, but other managers too, maintained a sensitivity to the detail of working conditions. A sense of personal security in knowing what to expect and a sense of identity were readily maintained by such procedures. Teleworking also used, as its core, a series of technological solutions which matched the employees' sense of themselves as high-tech people. Such methods seemed entirely appropriate for a company like Digital and the staff interviewed did not view them as new or awkward or technologically confusing. They expected to use high-tech solutions and would have found it unusual if the technology had not been sophisticated. It was not conventional at Digital for staff to bemoan their incompetence in the new high-tech world. Quite the contrary, even relatively junior staff could and did request training in complex up-to-date technological products, even though they might have only limited use for the services provided by those products. For example, R had begun her career as a secretary and was in her mid-fifties when the interviews took place. She defined herself as very non-technical, and was certainly the least technical person I spoke to in the company, but she took many technical things for granted: 'We've now got a training centre set up in Docklands which you can do the Microsoft like the Powerpoint, Word, Excel training and we're gradually, each of us are going on that because you have no back-up here, there's nobody, you know, in the next room, on the next floor to say "How do you do this?" So you really have to have a lot of knowledge on your own and be a lot more self-reliant. Oh very much so here. Very much so. I was mending the kit in the computer room downstairs the other day, phone in one hand, screwdriver in the other. It was quicker than sending out an engineer. And it worked, you know. In two minutes we'd got it up and running. It was a customer kit and we'd got it up and running again.'

Such demands were seen as normal. The sense of the company which came out of this was one where the staff interviewed saw themselves as skilled users of advanced technology and expected the company to be at the forefront of technology and associated working practices.

We thus find a complex, new operational routine incorporated into company norms. It involves working at home, with touch-down offices and advanced technology in spaces previously reserved for domestic purposes. The advice offered in, for example, the 'Flexible working handbook', and through formal and informal meetings, plus the provision of advanced technological fixes, mimicked the previous office environment. Some of this was deliberate, as the handbook made clear, and some emerged in use.

The new operational routine drew upon an interaction of structural features and features of agency. The company's market position was threatened, and this opened up opportunities for new working practices. Company structures were open to flexible working practices as a reflection of the product portfolio which had been developed in this area. However, the position of a number of strategic agents in relation to the power structures of the local organisation and the wider company, when added to their competences – and here the role of the senior manager was particularly important – meant that the potential solutions for the company were directed towards teleworking. The actions taken by these agents and those they influenced produced an outcome which had seemed far-fetched to some eyes but had become a natural solution. In enacting that solution the largely implicit need to maintain the security and sense of identity of staff produced particular practices, using technology in specific ways, such as VPN telephones, which tallied closely with previously adopted patterns of work, but not in other ways, such as computer conferencing, which were not part of the conventional pattern of working at DEC.

The new routine involved a taken-for-grantedness that many Digital employees worked from their homes and that they were provided with touch-down facilities when they used Digital locations. The complex technology which supported this was also taken for granted. (Right from the beginning of the fieldwork, conversations with Digital employees were always peppered with references to computer technology.) Personnel and management structures incorporated a formalisation of some hitherto informal activities, such as more focused meetings to provide business and social spaces, but informal contact also continued in new ways.

The practices which I have grouped together as a new operational routine were not entirely new. It is partly because they built upon expected and accepted values and enabled staff to maintain well-established and taken-for-granted ways of working, for example through the telephone network, that the move to teleworking succeeded. The discussions about teleworking from an early stage in the establishment of the telecentre recognised this, at least implicitly. The senior manager was determined to set up support networks which had 'to be endemic, it has to be seen as a thread throughout your organisation'. There is therefore an evolutionary process at work here. The particular routine adopted depended upon the interaction of structures, human agents and technology but it was built upon pre-existing patterns of working in a path-dependent way.

## The product portfolio

Given the adoption of teleworking as a conventional organisational form for its internal operations, it is somewhat surprising that the company abandoned flexible working practices in its product portfolio. Presumably, if Digital found such processes to be important as solutions to the problems it faced, other companies would too. Since the company had adopted teleworking itself, it would be easier for consultants to argue that customers might try the system for themselves. As an organisational form it was saving the company money and there was therefore a straightforward bottom-line argument which could be presented to potential customers. Why was the product portfolio changed in spite of the adoption of teleworking for internal organisational purposes?

A consideration of the two situations will clarify the factors which operated in creating or developing the operational routine. The company's review of the product portfolio was not routine but factors influencing it also influenced the changes in operational practices. The outcomes of the two events are apparently very different. Why is this?

Digital's market had changed through competition as new entrants came into the market and through consumer tastes which had shifted to personal computers and away from Digital's established products. The company had used teleworking effectively but was uncertain whether it presented a major new consultancy market. Its Alpha system of highly sophisticated, leading-edge computers was new and fitted into the experience and history of the company as a hardware-focused operation.

It became clear from the research interviews and the company's reports and brochures that the market was interpreted through this history. The decision to drop flexible working from the consultancy portfolio can be partly explained in terms of the company's taken-for-granted definition of itself as a manufacturer of, and support system for, electronic equipment. The new chief executive came from this background and must have been chosen in the light of the company's perception of its market place. Teleworking did not fit in with pre-existing notions of the nature of the company's products as manufactured computer systems and associated support. This is shown in the following commentary from M, the technical expert.

At the time of the interview, two months after C and S had been made redundant, M had been aware of the risky position of flexible working, as a product, for some time. He spoke of the dropping of the flexible working practices portfolio in a world-weary tone which implied that his pessimistic expectations had been fulfilled.

'The problem was that flexible working consultancy, particularly of the sort that [the senior manager] saw himself doing, which was talking to business managers about the advantages of flexible working, talking to managers about how you actually address the issues around changing to a flexible working environment, rather than "We've got these super Alpha boxes that mean you can build computer integrated telephony type services into something", were not, just not, seen as one of the key activities in terms of the overall company drive. The organisation that [C] and I worked in, up until late October last year, was called Digital Consulting. It was a consulting organisation and we were trying to move into high-end consultancy as well as the technology-end consultancy. We are now called Digital Systems Integration and we're much more technology focused than we are business focused. Business in terms of business consultancy. So the sort of services [C] was ideally set up to sell are not the sort of services, in terms of service portfolio, that fit naturally into what has been determined as our core business. So that's where the problem comes from.'

His interpretation of the new approach is done passively – 'what has been determined as our core business'. He claimed to be disaffected as a result of the changes which had taken place and this shows in his structuring of the argument. He appeared to be losing some of his ability or opportunity to influence the direction of the company. As the financial circumstances bit, there seemed to be a centralising tendency in management and a return to previously defined core business. The move back to the harder end of technology was confirmed by the change of name, from Digital Consultants to Digital Systems Integration.

B was a business consultant; that is, she was involved in looking for business solutions for her customers rather than solutions based solely in technology. She put it in a slightly different way. She was describing the company's preference for what she called 'selling tin', that is computer hardware, rather than the softer end of consultancy. She described how the company had moved towards consultancy: 'I think the norms have changed. There is a shift. I think there has been quite a major shift towards the consultancy area. And there is very much an acceptance of this is what it's all about.'

But the move to consultancy had been held back by the new-found success of the hardware. 'That potentially hasn't been helped, in one sense, by the fact that the Alpha systems have taken off in the way they've taken off. So almost from the move of an organisation from selling tin to selling consultancy, you know, they just happened to hit right with this.' The 'this' here referred to the Alpha system. Asked whether they were going back to selling tin, she replied: 'No they're not going back to it but it's, you know, for those people that have, are halfway there, so to speak, you know, they've got this to latch on, 'cos this is still, this is still going, and this is going well. But they have made a radical shift, as I see it. And there's this acceptance that this shift has occurred and it's what's got to occur. But if one would like to say the tin bit had been going for twenty-five years, the last bit's been going for five years and that I don't think we are expert enough at it yet.'

B's argument is, then, that the company had a set of norms which were to do with 'selling tin', and the softer consultancy services did not fit easily. In B's opinion, the future lay in consultancy and the company had made a 'radical shift' but the market success of the latest hardware had re-energised an area which she believed to be in decline relative to consultancy. There had been a shift back towards the hardware, arising from the historical preferences of the company over twenty-five years and from the initial success of the new Alpha systems. Business consultancy remained the way of the future but the focus was shifting towards consultancy linked to hardware (which meant, by implication, that flexible working as a product did not fit easily into the portfolio). It is to be expected that B, as a business consultant, would argue the importance of broader consultancy services. Her overall analysis is consistent with that of M, the technical expert, though less polarised. In her view, Digital, as a company, had a strong sense of its identity as a manufacturer and provider of associated support services.

B had been disillusioned with the general (though unspecified) management practices at Digital from the beginning of the fieldwork. She had found the support provided by the telecentre a major improvement, but the hiatus created by C's redundancy was a return to bad old ways. She did not comment on management structures, indeed she professed a healthy disdain for them. She did not regard changes in that area as important for her, provided she was given the support she needed. Her interpretation of events is based around the nature of the products on which the company concentrated: the company had moved to consultancy but was still not expert enough at it and retained a major interest in tin. M's reflections indicate disappointment and apparent loss of control over the direction he should take: '*we* were trying to move into high-end consultancy' but that is not 'what has been determined as our core business'.

The shift to more central control was confirmed by the senior manager. Formal relationships and processes, which had always existed but had not been exercised, had, in his experience, begun to dominate. The senior company hierarchy began to play a more significant part. C remained intensely loyal to Digital even after he had been made redundant but he expressed his disappointment at the way in which those formal procedures had been brought to bear. A year after he left he put it like this: 'The direction was very much from the top right down to the individual. Irrespective of, what, two layers of management between the Board and the individuals. And as a senior manager reporting into the Board I had absolutely no responsibility what, or accountability. I had a budget of, where was I, just before I packed my business in, about eleven million revenue, round about seven million expense and I couldn't sign an order for a box of floppy disks for my organisation. That had to go up to Board level. I use that as an example, it was that silly. How can you run? . . . This was basically, because of the state the company was in, its need to control expense. What I would rather have had was the MD tell me, [C], this is what you're working to. If you cross this line, [C], you're out the door. Yes. I don't mind you being that tough. I'd have rather have had that than not being able to manage my organisation. Therefore I think people having had years and years of their immediate managers giving them all this support and hand holding and encouragement and a lot of personal development, suddenly saw their managers, their immediate and next line managers, completely defused. . . . I think people inside still cared, people at Board level still respected others but Digital had a job to do and that was the way it was going to do it.'

C had worked for Digital for nineteen years and had technical and business expertise which gave him credibility. He had developed a network of contacts throughout the company and was regarded highly by them. He proudly displayed awards he had received in earlier work with the company as, for example, 'manager of the year', on his office walls. He had started working for Digital in their northern centre in Warrington. He came from the Warrington area and had the down-to-earth warmth and friendliness which is typical of Merseyside. He found it hard to criticise the company and always gave it the benefit of the doubt. A statement such as that above, showing his frustration at the loss of autonomy and control he had suffered, is significant. From around 1992–3, formal rules, it appeared, were being more rigorously imposed and the informal relationships, developed over years, were becoming less effective. The company's financial problems seemed to be dominating its espoused values. This was a marked change in management style, according to C and M, from that which had typified the company before 1992.

The company's *identity* based on a sense of its core business has parallels in the centralising moves reported by M and C. Centralising core procedures and focusing on core business can be interpreted as moving back to perceived core competences in an attempt to overcome the difficult market position in which the company found itself. Alternative strategies were available. The company could have redefined its business and concentrated upon developing broad consultancy skills instead of narrowing down. But B, the business consultant, was open about the company's relative weakness compared with mainstream consultancy firms. M's skills in technological developments were harnessed through a redefinition of his work as systems integration. As M indicated, this left little room for the flexible working portfolio. The company's sense of itself gave an edge to products rooted in manufacturing.

This need not have been so. Flexible working practices were in demand. R, as a team leader at the Digital telecentre, co-ordinating the work of tele-workers and providing clerical and administrative support for them, made the following observation eleven months after C had left the company: 'Well, unfortunately they decided to get rid of that group. Which again seemed an absolutely crazy idea to us in Newmarket because this is what we were doing. I'm still getting enquiries. Got one here yesterday. I think it was Dublin Airport wanted to try flexible working, teleworking and where do you go to? I ring up the two chaps I know who were in the flexible working group with C who don't actually do it officially any more but they usually try and help me out if I get enquiries and I do still get enquiries, at least one a month. "We hear you do teleworking." . . . Really there's quite a lot of interest everywhere. Seems such a shame to me when we've got these enquiries, you know, how do you do it, and we haven't got anyone there who can actually come out and sell the services to these people.'

R is a calm, unruffled person with many years' experience. She is very helpful and self-contained. Her frustration at what she saw as the silliness of the decision to drop flexible working was palpable. Many of the company's business leads in the East of England came through her and she was convinced that business opportunities existed. To her, the decision was inexplicable. This was not set up by her as a centre–periphery clash. C had been responsible for flexible working practices nationally, so, for that part of Digital's business, there was a sense in which Newmarket constituted part of the centre. There was a clear sense from her, however, that parts of senior management were out of touch with the potential of the product.

It is possible that the key decisions taken centrally did not address the detailed opportunities available for flexible working. This is certainly the view implied by M. Such opportunities did not figure in the response from a manufacturing/technological systems consultancy organisation. The orientation of such a company was based around hardware, not soft consultancy services. Flexible working practices were simply not part of the equation.

Furthermore, it can be argued that it is cheaper for the company to buy in expertise in flexible working practices directly from consultants and to move that transaction entirely into the market. Flexible working does not require highly specific assets. Staff can be trained in its use and the technology is readily available. However, given the ambivalence about the precise nature of Digital's market, the company could potentially have created a different environment had it decided to do so. The management skills and knowledge of human systems which C's group possessed are very difficult to replace. Certainly his colleagues believed this to be true. In these circumstances the company could have become a market leader in flexible working practices by valuing the synergy of its skills in human systems and its competence in providing sophisticated technological solutions. In developing flexible working the company would have changed the nature of its environment and particularly of itself as a company. This decision was not taken.

According to a Digital press release (2 April 1997) 'Digital Equipment Corporation is a world leader in open client-server solutions from personal computing to integrated world-wide information systems. Digital's scaleable Alpha and Intel platforms, storage, networking, software and services, together with industry-focused solutions from business partners, help organisations compete and win in today's global marketplace.' Its focus was clearly still on business solutions provided through specific hardware and software, on 'selling tin'.

The decision to drop flexible working from the product portfolio, while incomprehensible to R and regretted despairingly by M, can be explained by the way in which the firm interpreted the market and its place within it. The company was defined by the new CEO and by the way it was structured as one focused on hardware. At a strategic level beyond the participatory reach of the Newmarket operation, flexible working consultancy services were not of major significance. Market pressures forced these meanings and a harder interpretation of the structures to the surface.

To an outsider this could be interpreted as conventional behaviour under economic pressure. Inside Digital it was not so straightforward and we need to understand further the culture at Digital to understand why this was so.

## The culture at Digital

The way in which the company decided what it should do has been set up here as a debate between different sets of meanings about what it was, or what it should become. The routine adopted for operational organisational purposes at Newmarket was built upon a particular set of staff attitudes and beliefs about the company that was to some extent at odds with the moves taken nationally and internationally with respect to product strategies. The use of teleworking in the company, in the way in which it was set up at Newmarket, as I have shown, matched and mimicked existing routines and the staff's sense that the company was serious about its founding values. This was not incompatible with a simultaneous interpretation of the market that saw teleworking as an inappropriate product, though clearly the Newmarket staff had a different interpretation of market potential and appropriateness.

The Annual Report in 1993 claimed the company believed in the core values which had made it great: integrity, valuing individuals and their diversity, fiscal conservatism, innovation and technical excellence (Digital 1993). As already noted, it had been established by Ken Olsen in the tradition of Quaker businesses. The staff interviewed at Newmarket signed up to these values.

M, for example, in my first meeting with him in early January 1995, described the company's approach to team building in this way: 'In terms of the way we deliver to our customers we will have a project team. The project team [will be] . . . bringing in the right technical resources, delivering resources, administrative resources and at that level we will have a great deal of flexibility about how the work is centred around those teams. Bringing them in not simply from our own local Professional Service Centre but just looking around, I suppose the UK in most instances, but if necessary we will go off to Europe and bring in expertise from Europe'.

M refers to the company as 'we'. This was common and was used with a regularity which gave a clear message that those interviewed identified themselves as part of the organisation. He describes how he has considerable autonomy as a senior technical expert and how *his* team can look for the best people. (The Professional Service Centre is the Digital term for the local centre to which consultants are attached.)

Even after the blow of the winding up of the flexible working practices team, M still expressed a commitment to what he saw as the company's values: 'but let me step back a little bit and say when I joined Digital – which is now coming on for eight years ago – the thing that impressed me about Digital was the way it very clearly cared about people. And I guess what is true is that there was a feeling that if you cared about the people then the business looked after itself. Clearly then, have run through a period recently where it became clear that that had not worked. The business was not looking after itself and there was therefore much more focus on the business. There was never a pull-back, deliberately, in terms of people management. If you look at the statements coming out from the top of the company when Bob Palmer took over Ken Olsen's role, one of the first statements he made was to reinforce all the messages from Ken, both in terms of importance of people within the organisation and also in terms of the ethics of the organisation which,

I don't know if you're aware, but Ken Olsen is a Quaker and he was not in this business to make money. You know he made a lot of money, he was a very successful businessman, but in a sense what drove Ken was technology and service to the community, both to the Digital community and also the wider community, the customer community and a lot of Ken's money has gone out to charities rather than to buying himself the latest, biggest, fastest car. He drives around in a ten-year-old car apparently and things like that.'

This is an interesting reflection. M is confirming the culture as a caring company. His use of the familiar first-name address for Ken Olsen did not come across as an attempt to impress or in an awkward way. He did not see a 'pull back' from the initial care but conceded that the approach of allowing the business to look after itself would not work in the changed environment. In other conversations, he was bitter about the treatment of C but saw it as a mistaken policy rather than necessarily a shift of company culture, though he did see some shifts occurring in the culture. There is an ambivalence here. The company culture was strongly embedded. The evidence, of which M was clearly aware, could be used to demonstrate a shift to a much harder-nosed culture but M resisted that interpretation.

In the period when the telecentre was becoming established, C referred to his own role with a sense of ownership of the project. He shows that ideas from individuals could be incorporated into the business and has an approach to management which is facilitative: 'Going to blow my own trumpet here. I guess, I don't know, it's difficult to say, I'm not sure there were many people who believed in flexible working at the time. I think it happened that I was probably one of the more mobile workers in that sense . . . I put forward some ideas, I obviously try as a manager, try to demonstrate some leadership. I was keen to pursue flexible working as one of the options.'

B was more cynical about the company as an employer but her language also adopts an inclusive 'we' approach. Here she is, interviewed two months after C was made redundant, describing the approach to consultancy at Digital: 'And I don't think, and I'm not just saying at Digital, I think it's a general, sort of, computer company, all the computer companies moving across to consultancies in this area, I don't believe we're experts to the degree that we need to be out in the market yet. We're not consultant experts like you would put down against a Price Waterhouse or an Andersens yet purely from the, shall we say, the whole consultancy approach bit. We're moving there, we're getting better all the time. As I say not just us, all the people like us that have taken this shift on board and are going for it. But that's how I see it. We're not ruthless enough yet, almost.'

B gives a sense of the autonomy and search for an approach which was shared in different ways by the consultants at Newmarket. Staff took respon-sibility for their own actions. Such acceptance – in many respects, welcoming – of responsibility carried through to the co-ordinator/team leader roles. S referred to her work in the same kind of language. In the quotation below, she describes the way in which the telecentre was developing in its early period. She shows team working, the use of 'we' again, meaning that different staff took actions using their own initiative, alongside a realism about what can be done. The interview took place early in the fieldwork when the telecentre had been open about six months: 'I think maybe at the beginning there were teething problems because we weren't following an example, [C] was doing this for the first time, so it was live and going ahead for the first time, we had no book, no guide book, there were no rules to follow. [C] was making up his own rules which worked quite well, but obviously there were issues that came up, so if we were going to open a telecentre elsewhere we'd know to deal with these things up front, for instance the telephonist/receptionist area, if we opened a fairly large telecentre again, rather than putting the phone calls across to someone else we would suggest now that there is somebody installed who could answer telephones and to do personal duties and perhaps facilities.'

The company culture at this time, in the early months of the telecentre, can be seen to be positive about new ideas. Staff took responsibility and showed initiative. The claims made by the company and quoted earlier that ideas 'bubbled up through the entire company' were not wholly rhetorical. There was a belief that flexible working practices could save the company from its financial plight and release staff from the drudgery of commuting. Furthermore, the success internally was seen as advantageous in selling such practices through consultancy, where more revenue could be generated, thus creating greater security. The fear that the Newmarket office would close had influenced such attitudes. C was seen as a manager who, with others, had saved the jobs of his colleagues. There was a cautious, positive approach tinged with uncertainty about the future. There were very positive attitudes to local management but less confidence in the understanding of the Board.

C was a champion of the teleworking idea and his enthusiasm was infectious, but other staff readily volunteered their own belief in the scheme's effectiveness. The important point here, however, is the positive culture and the meanings which staff used to make sense of their circumstances, which has to be placed alongside other formative events and structures in tracing through the way in which routines evolved at Digital.

In the early days of the telecentre, therefore, it is possible to observe a set of meanings which emphasise opportunity and initiative. As the quotations show, staff in Newmarket became less confident about their ability to change things. Both M and C, however, retained a view of the company as an organisation which cared for its staff. At the same time they and others interviewed saw a different but closely related set of meanings being imposed as financial constraints bit more deeply and the company began to focus on a different, and contested, interpretation of its core business.

An element of cultural conflict can be seen. A particular set of practices had been established at Newmarket. This operational routine was built on previous practices in a path-dependent way and took for granted a culture which would be supportive of staff and give them autonomy to manage their own work. Within that framework difficult changes in working practices had been put in place. The national and international levels of decision making, however, viewed the company differently: the operational changes at Newmarket saved expense. They were important and helped to provide the opportunity for the company to focus on established core product categories.

With such distinctions in meaning and focus it is possible to reconcile the potentially incompatible actions of establishing teleworking as a conventional mode of operating and dropping flexible working from the company's product portfolio.

## Did technology drive the changes?

Technology partly defined the possible and was also a symbolic part of the company's routines. The use and sale of technologically sophisticated machines was Digital's core business. Internally, the relationship between technology, structure and agency had positive feedbacks. Staff and the company knew about technology and used it in complex ways – confirmed by teleworking. It was considered desirable to be at the forefront of such developments. Orlikowski's (1992) analysis using a structuration-theoretical approach maps closely onto the events at Digital. The technology was clearly the outcome of design and development activities and facilitated and constrained human action. The routine of flexible working influenced human interaction with the technology. The institutional properties of Digital then changed as new teleworking practices were incorporated.

All staff were very conversant with technical systems. C insisted that technology was enabling, not determining, as a factor in setting up new working practices, but he spent a high proportion of any explanation of the system describing the technological wizardry which made it possible. The technology was taken for granted as a basic framework, just as the availability of paper and pencils might be viewed in a less technically sophisticated clerical environment. Without it things would not be the same, but all respondents were clear that the technology had not promoted the drive to flexible working. It had permitted it.

On the product side, technology was a basic element and the meanings ascribed to it were important. Supplying business services was not deeply technological in the sense in which the term was understood at Digital. Service came as a part of a package supporting technologies (seen as artefacts). The approach to business consultancy typical in the consultancy industry, in contrast to the computer-manufacturing industry, was foreign to Digital (and claimed to be foreign, by Digital staff, to other computer manufacturers). A particular interpretation of technology was recognised at Digital and this fed back into further interpretations of what constituted legitimate activity. Selling soft services touched only the edges of that interpretation.

Thus we can see that an analysis of the complex interaction between institutional (and other) structures, human agency *and* technology is required in order to fully comprehend the nature of the events at Digital. But the technology does not exist as a separate entity. It is implicated in and constitutive of the way in which the company is understood, the way in which the staff view themselves and the way in which they define what is and is not possible. Particular interpretations, exemplified by teleworking, view technology as permissive. The interpretation taken by C, for example, fails to recognise how deeply technology was embedded in the history and culture of the company. The technology did 'enable' but it was also part of the network of activities which created the opportunities for teleworking in the first place. At Digital the technology was an unquestioned part of the world and had an intimate part in the recursive relationships between structures and human agents which were instrumental in the changes which took place. The technology was not transformatory in itself, however. In that sense it did enable change to take place rather than drive the change.

## In summary

A new operational routine – flexible working – was developed and created when:

- the market was changing unfavourably for Digital; its technical excellence was no longer sufficient to mark it out from the competition and it was overstaffed by industry norms. Thus the company engaged in a search for solutions;
- the new routine was compatible with the firm's culture about the kind of activities in which it should engage;
- existing rules and other features of structures were similar or could be mimicked in order to provide a sense of security, identity and continuity;
- staff suggesting or enacting particular new forms of behaviour were in a position in the company where they were able to influence the agenda about appropriate responses to changes. This influence came about through actions and working practices which were considered appropriate as well as through deliberate argument and persuasion;
- the skills of staff were aligned with the proposed solutions;
- staff facing changes could identify with the proposed solutions in ways which confirmed their own sense of themselves and the company;
- the changes built upon pre-existing practices.

There was a resistance to change in the company when:

- activities did not fit with the company's sense of its identity in terms of the market in which it was engaged;

- well-established company practices were incompatible with the new routines;
- strategic agents were not significantly involved in the development of new products.

Technological excellence underpinned all activities practically – the technological fixes were needed to make the system work – and conceptually as the-way-things-are-done at Digital.

These factors operated simultaneously and acted back on each other and on themselves. Without any one of them it is likely that circumstances would have turned out differently. The East of England operation ultimately had relatively little control over the interpretations of the market place made by the wider company. It was not within their remit. Their successes as tele-workers were recognised but this fed back into the company as a valuable structural change which was financially sound. It matched the company's high-tech, creative culture and was self-reinforcing. At the product level the feedback loops worked in the opposite direction. Consultancy on flexible working required little hardware. It was labour-intensive at a time when the company saw itself as overstaffed. Any single consultancy of this kind generated relatively small revenues and, furthermore, the consultancy market was occupied by companies operating to different recipes who seemed more experienced and powerful at this kind of thing.

Strategically, the agents operating at the level of decisions about the company's product mix were not involved in detailed evaluation of the success of flexible working. Their norms were related to large, technically sophisticated, hardware solutions. In terms of organisational structure in the East of England, and ultimately much more widely within the UK operation, there were agents like C and M, who operated in ways which changed strategies. Their formal presentations and rule changes, and informal routine behaviour, generated change. In decisions about product mix, taken much further away from local activity, their impact was felt much less strongly.

In trying to understand the changes at Digital it has been necessary to look at the processes of change and the ways in which they were interrelated. This has involved a consideration of the factors which were lying behind the surface phenomena. Underlying structures can be found at a level beyond the firm, that is relating to features of the market. They can also be seen at the level of the firm itself in the relations between the quasi-autonomous Newmarket centre and the larger company within the UK and world-wide, and in the way in which staff responsibilities and relationships were structured. Management support and the rule requiring managers also to carry out direct revenue-earning activities were important structural features, for example, influencing the direction of change at Digital. Without this particular rule it is possible – indeed likely – that C would not have been made redundant. Finally at a structural level, conventional views of the boundaries of the firm were challenged as people's homes became (again) a legitimate location for work.

The developing routine evolved. It developed partly through mimicking existing structures and relationships while changing them in small ways, and in this case it was largely deliberate mimicking. Rules were suggested and adopted which became habits (Hodgson 1997). The success of teleworking as a process within the firm partly depended upon its previous success as a product: this is an irony and shows how the search for new routines depended upon structures, agents and knowledge already in place. The new routine can be seen as a mutation with some deliberate and less deliberate elements. Other factors (external to Newmarket and the flexible working teams) changed the way in which teleworking was viewed as a product. The change was gradual, finances were tightened, different staff were made redundant. A product strategy *emerged* though in some ways it manifested itself as a quantum jump (Miller and Friesen 1980; Mintzberg and Quinn 1991) when the flexible working team was ultimately wound up. An evolutionary, path-dependent approach shows how changes depended upon the structures and practices which were already in place, as well as on the activities of particular agents. The (partly) chance occurrence of a group of agents skilled in teleworking and the pressures on the company for change resulted in the development of a new routine which was reproduced elsewhere in the company and survived in the company's practices and memory after key strategic agents had gone. What is clear is that the operational practices evolved through the interaction of different levels of structure, agency and technology which was interpreted through a particular company culture and founded in a particular history.

# 5 The National Extension College: strategic routines

## Introduction

It is frequently an implicit assumption in organisational analysis that routine behaviour is primarily, if not exclusively, operational. The seminal work of March and Simon, referred to in Chapter 2, is explicit and discusses a continuum which describes behaviour moving from completely routinised to 'problem-solving' (March and Simon 1958). This distinction is too stark in arguing that one end of the continuum is routine and the other not routine. In this chapter the concept of a *strategic* routine will be developed from the observations and interpretations of events at the National Extension College. The implication is not that strategic decisions are automatic but that the range of decisions is limited and that the strategic processes, of searching for better ways of doing things or better things to do, are confined within a narrow band by the strategic routine(s). This argument, therefore, rejects the view that routineness is only an operational feature. It claims that there are established and sanctioned practices at strategic levels which are the taken-for-granted ways of doing things. Furthermore, operational routines have implications for strategy so that routine forms of behaviour frequently contain activities which are based in day-to-day operations and at the same time have a direct bearing on strategic decisions. This is the key point: organisations possess strategic routines. The argument is not that all strategic decisions are routine but that routine processes significantly affect strategic levels in organisations.

The chapter concentrates on identifying and analysing the strategic routine adopted at the NEC and how it influences and is influenced by both the internal and external worlds of the College. Strategy is about the organisation's relationship with its external environment and is here taken to refer to *the organisation's attempts to shape its internal and external environments so that it is better placed to achieve its objectives or to redefine its objectives*. Strategic routines significantly affect these environments but the routine is not necessarily deliberative. Taken-for-granted behaviour which is carried out skilfully without meriting comment by the organisation has important strategic dimensions.

In addition, one part of understanding the development and reproduction of strategic routines is to consider the extent to which they fit with the behaviour of suppliers, customers and other associates of the company. This idea of strategic fit will also be used to analyse routine patterns at the NEC.

The change considered in this research is related to technological developments. A primary research issue is to understand the extent to which technological change *drives* other changes or *enables* change to take place whose primary causes lie elsewhere. In the search for suitable fieldwork sites, it was important to study an organisation which was *beginning* to use new technologies for the first time on the grounds that if the new technologies make any difference then for new users the potential for change would be high. To find such a fieldwork site requires a good deal of luck. It is not easy to discover which organisations are *planning* to use new technologies. Organisations either do or do not use them. The opportunity to use the National Extension College as a fieldwork site occurred through a chance meeting with its Director. During a social occasion she expressed an interest in the project and subsequently agreed to participate. A decision had been made at the NEC to implement a new information system a few months before the fieldwork began. The fieldwork took place during the period in which the information system was being introduced and it was possible to look for the ways in which it was implicated in changes at the College.

What follows then is an identification of the strategic routine adopted at the NEC, viewed partly through its interaction with the new information system. The argument is that the behaviour of the College cannot be fully understood without an analysis of its routines and that the routines themselves, and the factors which operate to reproduce them, are accessible through an interpretation of the actions and discourses of members of the organisation.

## *Background to the College*

The National Extension College was the brain child of Michael Young (latterly Lord Young of Dartington). He announced its establishment in the October 1963 edition of the Advisory Council for Education (ACE) journal *Where?*, and clearly had in mind a distance-learning organisation based around the use of new technologies. It was conceived as a distinctive, radical and optimistic attempt to provide educational opportunities to those who had been denied them. It was a clear forerunner of the Open University (in which Michael Young also played a part).

Young's original conception was less technologically radical, however. He envisaged fitting a second Cambridge University, offering higher education to working-class and mature students, into the five-month vacation of the first but, after visiting the Soviet Union and observing the use of correspondence tuition there, he modified his ideas and set up the NEC in 1963 using grant and loan money from a number of charitable foundations and individual contributions. In reviewing the College's first twenty-five years in 1988, Young said,

> An institution is the way a habit perpetuates itself. That, or something like it, is certainly true of many institutions: they do tend to ossify. They suffer from a hardening of the categories. But the National Extension College is most unusual. It was born by enterprise out of zest, and it has stayed that way throughout all of its 25 years. . . . This new era will need more than ever before NEC's gift for flexibility, innovation and openness. . . . The ethic is that of public service, even though the means by which this ethic is given expression is non-profit private enterprise.
>
> (National Extension College 1990)

Young's comments give a clear sense of the way in which those interviewed and contributors to the twenty-fifth anniversary publication (to which the above quotation is part of the Foreword), view the College. They also indicate the kind of culture that has developed. The title of the anniversary publication, *The National Extension College: A Catalyst for Educational Change*, emphasises also the innovative and flexible approach espoused there. There has been some 'hardening of the categories', despite Young's claims, as will be discussed below, but also considerable change, and recently some softening of previously secure categories.

As Table 5.1 shows, in 1994, when the fieldwork began, the College had a turnover of over £3.5 million with a net profit close to £400,000 (a profit margin of 10.72 per cent) and it employed over fifty people. It was going through one of its most secure and successful phases. It had not always been so. Despite Michael Young's optimistic words the College has faced regular crises. Its finances have been insecure and its focus imprecise as it moved from one financial opportunity to another. It has retained its underlying mission of using distance-learning techniques to do its best for students who are educationally underprivileged. And in searching for revenue and financial support and developing new ways of providing educational opportunity, the College has necessarily found itself part of a network of writers, tutors, other educational establishments, broadcasters and others.

Such a style of working goes back to the beginning. An inspection report commissioned by the NEC from John Blackie, a retired school inspector, in 1970 described a small number of core staff struggling under difficulties and linked to a wide network of tutors and course writers (Blackie 1970). Partnership with Channel 4 and other television companies during the 1980s was a further manifestation of this style of working. Similarly, at the end of the 1980s the NEC was the largest single provider of Open College courses. Richard Freeman, the NEC's Educational Director from 1972 and Executive Director from 1976 to 1987, commented in 1983 that such partnerships were a strength of the institution. Looking back to its early years, he said, 'Here was a college prepared to tackle things which at the time were risky, like

*Table 5.1* National Extension College Trust Ltd, company financial profile during the period of the fieldwork

| | 12/96 | 12/95 | 12/94 | 12/93 | 12/92 | 3- or 5-year average |
|---|---|---|---|---|---|---|
| Turnover | not available | not available | 3,620,094 | 3,579,430 | 3,349,848 | 3,516,457 |
| Profit before tax | 275,510 | 617,755 | 388,138 | 179,457 | 249,144 | 342,001 |
| Net tangible assets | 2,389,302 | 2,113,792 | 1,496,037 | 1,107,899 | 928,442 | 1,607,094 |
| Shareholder funds | 2,389,302 | 2,113,792 | 1,496,037 | 1,107,899 | 928,442 | 1,607,094 |
| Profit margin (%) | not available | not available | 10.72 | 5.01 | 7.44 | 7.72 |
| % return on shareholder funds | 11.53 | 29.22 | 25.94 | 16.20 | 26.83 | 21.94 |
| % return on capital employed | 11.53 | 29.22 | 25.94 | 16.20 | 26.83 | 21.94 |
| Number of employees | 71 | 66 | 51 | 53 | 47 | 50 |

Source: FAME Financial Analysis Made Easy Database.

Note
All figures are given in £ sterling except where stated; columns refer to the financial year ending in the month indicated.

training social workers at a distance or teaching people to build radios in their own homes. Yet it had virtually no staff or resources and it could only succeed if it could persuade other institutions to change. All its major experiments were collaborative – with ITV, with BBC, with WEA and so on. By itself NEC could provide very little but acting as a catalyst, it triggered a major extension to education for adults' (Freeman 1983, quoted in National Extension College 1990).

Currently the NEC continues to exhibit features of this kind. Many staff have worked for and with the College for many years and carry the College's frames of reference with them. New members of staff are recruited from a population familiar with and sympathetic to the College's aims and methods (interview with College Director 1994).

The NEC has been subjected to a number of changes during its existence primarily as a result of its need to find funding for its projects. In organisational terms, however, there appears to have been considerable stability over a relatively long period. It has remained, at heart, an *educational* institution serving a client group who for one reason or another have been educationally disadvantaged. It has been radical in the soft sense of providing courses and materials by relatively innovative methods but has not set out to change the world. Rather, it has been pragmatic and opportunistic in finding activities which generate sufficient funds for it to carry on its core business (National Extension College 1990; interview with College Director 1994; National Extension College 1996).

During the period of the fieldwork the College was undergoing two unusual (to the College) changes. The first was the establishment of the information system. The College had used information technology, broadly defined, during the whole of its existence. The original 1963 conception had emphasised such matters. But this was the first time that a comprehensive approach to information input and retrieval had been attempted.

The second unusual change was the retirement of the College Treasurer. He had been Treasurer of the College from 1965, almost from its inception. As a measure of his influence and the affection in which he was held, a photograph of him was hung in the College entrance hall, subsequent to his leaving, alongside those of Michael Young and Brian Jackson (the College's influential first Director). No other portraits are displayed in the College. The Treasurer had enormous influence and took on a wide range of activities, some of which bore only a distant connection to the conventional work of a treasurer. The Treasurer's retirement had an impact upon the behaviour of the College and this will be referred to again since it symbolises some features of the College's culture.

## Staff directly participating in the research at the NEC

Five NEC staff participated in the research:

O:  the Director of the College, who reported to the College trustees and, when the fieldwork began, had worked at the NEC in various capacities for a considerable period of time;

L:  one of four Assistant Directors reporting to O, he was responsible for educational development and had been at the College for five years;

T:  at the NEC for eighteen months, she was a co-ordinator for tutor appointments, supporting, monitoring and recruiting tutors for courses in all areas of the College other than degree and professional programmes;

E:  an IT specialist who had worked at the NEC for over two years, he took day-to-day responsibility for all IT activities and advised the senior management team on policy issues; midway through the fieldwork E resigned from what was becoming an increasingly managerial role and remained at the NEC as an IT consultant;

A:  A took over as IT manager on E's resignation, and was new to the NEC; at this time the IT department had grown from E on his own to a unit of six people drawn together from other parts of the organisation.

## The strategic routine

The NEC's primary strategic routine can be identified as operating through networks and partnerships. That is the way in which things are carried out in dealing with big changes, with suppliers or customers or in internal organisational matters, regardless of the appropriateness or detail of the approach. The routine is consistent with the opportunistic and pragmatic style of the College (Blackie 1970; Freeman 1983) which, in turn, was partly built upon the need to look for revenue-earning opportunities. Justification for this claim is developed below.

The routine also appears to have a good fit with the College's suppliers and customers. Writers, tutors, training departments, students, educational professionals, for example, find it easy to fit into the structures developed at the College (interviews with Assistant Director 1995, 1996; interviews with College Director 1994, 1995, 1996). And the College, in the sense of the collectivity of staff who work there, finds it relatively easy to respond quickly to the demands of suppliers and customers. The College defines its areas of interest partly through its methods of operating. If new business is acquired the routine is modified in subtle ways to meet the new demands but the underlying approach remains rooted in networks and partnerships. Below are some examples of the routine in operation.

O regards the ability to respond to potential partners as one of the College's strengths. Asked whether they had to respond very fast, she answered: 'Very, very fast, yes yes. And that is, if we're looking at market strengths I think that is one of NEC's assets. That, I mean we have had, we've worked in partnership with a large number of organisations. I mean we're quite an acceptable partner because we're non-profit making, we're an educational trust so quite a lot of people want to work with us.'

This is partly because the College does not have the resources to fund all its development activities. 'That we know that if we're going to get into that area seriously we'll have to do it with partners 'cos we haven't got the sort of investment capital that you need to go seriously into any multi-media initiative.'

There are many potential partners and, as a result, O spends much of her time acquiring and developing networks. 'I can't tell you how much time I spend. I suppose I spend at least a part of every day, unless it's blocked out to do something completely different. And it isn't actually very time consuming. But the reason why I do it is because it gives me a basis which helps me make decisions about NEC. . . . if we are approached by an organisation, an individual that I've never come across, I will do my very best to find out as much about them as I possibly can, and I will use a network of friends, colleagues, ex-colleagues, people who I've bumped into at a conference, gave me their card, I'll use it, you know. And also expect people to use me in the same way. I mean you know, I don't, I hope it's not just a one way thing. But I would reciprocate always.'

The College Director thus bases her strategic approach around the acquisition and nurturing of networks and partnerships. It gives her the 'basis which helps me make decisions about NEC' and she will 'reciprocate always'.

Other senior staff find themselves operating in similar ways. The College sees itself as a partner and collaborator. Internally, similar attitudes are adopted. Homeworking for staff has been a feature for many years. L said: 'It's slightly complex because some people work on a contract basis in and out of house and some people are freelance but work on a very regular basis with us. In terms of direct sort of full-time staff I think about seven at the moment, but there is a group of what we call project managers who are out of house to develop new publications. . . . They are freelance people, but they work probably at least 50 per cent of their time with NEC projects. And so we consider them very much as part of the organisation and try to make them feel involved as much as possible.'

L's uncertainty about precise numbers and the precise contractual relationship of his staff reflects the way in which individuals are seen at the College. There is no clear sense of core and non-core. Individuals are judged according to unwritten criteria.

Internally partnership arrangements are seen as appropriate by those involved in strategic decisions. This is not simply a hands-off form of subcontracting. The freelance staff are 'involved as much as possible' in the organisation. The NEC also uses other freelance staff in a more arm's length relationship. Such staff are usually sub-contracted by the project managers (who themselves are usually freelance) and in those cases the concerns of the

College are simply to ensure that the finished product is up to standard. However, once members of staff are perceived as partners and colleagues they are treated accordingly. It is difficult to know precisely how this boundary is crossed but once it has been crossed individuals appear to hold onto that status. For example, though E, the IT specialist, resigned from his full-time post and worked only as a consultant he was referred to by other participants, and apparently treated, as a full member of the College. His new role was as a 'consultant' but his tasks were to deal with a full range of issues which in many institutions would be picked up by staff on permanent contracts. This style of relationship was not worthy of comment at NEC. It was taken for granted.

This strategic routine is embedded in the institution at different levels. The history of the NEC, both in very local terms – its founding fathers (Michael Young and others) and long-serving staff – and in terms of the optimistic philosophies of the 1960s from which it sprang, has given it a strongly held set of ideals through which it views the world. Thus it continues to view itself as an *educational* institution and in 1996–7, for example, *reduced* its prices in order to share its new-found security with its clients. This is a remarkable step to take. The clients thus appear to be seen, at least implicitly, as partners. The strategic routine is to share throughout the College network. The sharing, in this example, took a very practical approach.

The College thus gathers intelligence about its environment and develops relationships with other organisations by operating through partnerships and networks. The approach to staff relationships has the same broad pattern and the College's view of its clients or customers is also based on the same kind of activity. Decisions which are taken and processes which are adopted use this style. This is not simply a view of the world, it affects the ways things are done.

No other strategic pattern springs readily from the data. It would be inconceivable, for example, for the Director to make a unilateral or impulsive decision. Broad policy matters are debated. The concept of routine used here does not preclude that. The debates are conducted in staff meetings and with the College's trustees where the same approaches are reported. The trustees are unpaid and drawn from the College's networks. (For example, at the time of the fieldwork one trustee was an educationist who has been among other things a senior school inspector, another had risen through the educational inspectorate and was then Principal of a Cambridge college, a third was an ex-Cabinet Secretary, Cambridge academic and one-time Pro-Chancellor of the Open University, another was the chair of a national adult education charity, another Director of Continuing Education at Cambridge University.) This is not therefore a surprising outcome. It would be surprising if it were otherwise: this indicates how thoroughly this style of working imbues strategic practices at the NEC.

Routines are not unchallenged, nor interpreted identically by all those connected with the College, however. We need to consider where the routine comes from and to examine examples where such forms of behaviour are challenged and how those challenges are dealt with.

## What meanings lie behind the strategic routine?

A pragmatic and opportunistic approach has been adopted in developing the College's portfolio of projects (National Extension College 1990). The College's history has been characterised by financial insecurity and uncertainty about the projects and courses in which it is involved. The NEC has to watch developments in its area of interest closely in order to ensure its survival. The College Director was very happy to accept that the College was opportunistic (in the conventional, not Williamson (1975), sense). Such a term did not misrepresent it: 'No, not at all, no. Extremely opportunistic. And that has come from I mean, I suppose the reality of staff knowing that if we don't bring in enough money their jobs are on the line.'

But the opportunism is tempered with an approach which selects those projects the staff feel comfortable with. There is a level of pragmatism and flexibility which is also based on an implicit understanding of what the College can do well and believe in. The College Director described the rejection of a potential money-spinner: 'I mean just to give you one concrete example of that, for a number of years we ran a telecoms programme for BT which was, had a BTEC accreditation and it gradually wound down, mainly because of the privatisation of BT and the qualifications were no longer a prerequisite for promotion at technician level. . . . Well just very recently we had a meeting with City & Guilds who also had telecoms qualifications and the people from City & Guilds said "We always wondered why you stuck with BTEC for those qualifications, as ours are, you know, much bigger, very much faster and it's still one of our fastest-growing areas with a tremendous amount of take-up overseas as well, as countries try to get their telecoms sort of expanding around the world" and they said "Why you don't you go back to Telecom, you know you'll make a lot of money out of it." And after he'd gone we sat down and talked about it and there is honestly no one in the organisation who is in any way interested in developing a whole new telecoms scheme.'

Thus the approach involves an assessment of the potential revenue from any project and the extent to which the project fits in with preferences at the College. The need for an individual to take responsibility and drive projects forward is recognised. But projects do not arise only by chance. The culture is opportunistic in picking up things which happen to fall the College's way and the underlying sense of what kinds of activities excite NEC people makes that opportunism partial. But the perceived market place is watched carefully, which creates a level of intelligence about what is likely to arise. The College Director again: 'Well from, if we look at other distance-learning providers, we will many, many times approach them as potential students. You know, members of staff will do that, will get on mailing lists, will, we will analyse the offerings and compare it with what we're offering. And that means that our own customer services team, when they're talking to students, can actually talk to them in a knowledgeable way. With publishers we obviously get the catalogues and compare as well. There's a lot of networks and grapevines. I mean I think one of the interesting phenomena over the last few years has been the number of people who are working on a freelance basis, and are working for a number of organisations in similar fields, and quite often through those sorts of grapevines we hear that, you know, another organisation is planning a competing product to ours. We, I mean one thing I do personally, is I go through all the job ads in the education press, not because I'm personally looking for jobs but quite often it's the first announcement you get that somebody is planning a particular initiative, or planning to move into a particular sector, because they advertise for staff in that, you know, to set up a new exciting unit to look at the Health Sector or something like that. And I mean within the organisation we've all got our antennae out and we pass information around.'

Collaboration in 'networks and grapevines' and the historical tradition of employing freelance staff figure widely in this comment from O. It shows how central such relationships are for her way of working and for the work of many other people at the College.

Obligation and reciprocity are important features of the NEC culture. The sense of reciprocity is largely unspoken but shows in many aspects of College life. For example, a ritual which confirms aspects of that kind of meaning is the way in which staff there celebrate birthdays through buying cakes for the rest of the staff. Interestingly, this was drawn to my attention because it had become too extensive. As staff numbers increased cakes were being provided so often and in such quantities that a new convention was agreed, that there would be a day set aside every few weeks when several people would provide the rest of their colleagues with goodies. This was presented as a minor aside to illustrate the growth in staff at the College, but it also reveals the depth of the culture and helps us to understand the meanings which NEC staff used in interpreting events. The obligation and reciprocity which the NEC as an institution drew on in almost all its relationships is encapsulated in this small ritual. Such meanings were then carried into taken-for-granted ways of operating at the College.

Alongside the care taken with personal relationships, NEC staff are flexible and used to adopting new methods. E was surprised at the acceptance of e-mail: 'I mean I was surprised when I put it in, it was the second e-mail system that I put in, how widely accepted it was. They'd obviously had some, I've noticed with e-mail systems when you put them in there's a sort of, the first phase is, sort of, novelty value, people send silly messages to each other just for the hell of it, just playing around, but there didn't seem to be that much of that here. People seemed to take it very seriously right from the word go. And I mean, certainly, if I look at the usage on the server it's surprising how many messages people are sending.'

However, surrounding the espoused flexibility and openness is a sense of tradition which can sometimes appear stifling. The Treasurer emerges in the accounts given as a man of enormous influence. His style appears to have possessed elements of paternalism. The College Director spoke on a number of occasions about the Treasurer's notion of the 'little people': 'He's got a wonderful sort of turn of phrase and he distinguishes between the little people, who are the freelance people or people who run very small businesses, and the big people, who are British Telecom and Parcel Force and Inland Revenue and people like that. And the cash flow, for instance, he obviously is, you know, ensures that the big people get paid on the last possible date, he doesn't go over but, you know, he holds back the cheques. But the little people he will pay on the dot and it makes an enormous difference, I can tell you.'

The enormous difference, according to O, was the loyalty shown by the little people to the NEC. In general it was claimed that they preferred to work for the NEC and would deliver work for it ahead of that for other organisations because it treated them well. They appear to have been drawn into the NEC network.

I had a sense of the Treasurer as a strongly paternal figure and this image was not rejected when I suggested it to the interviewees. Indeed, several respondents positively confirmed it, but it did not appear to be resented or resisted. He was a 'fabulous bloke' (T, the tutor-co-ordinator). The cultural norms which he personified, including elements of 'father knows best', were not entirely consistent with networking, however.

There is an element of contradiction here. Networking brings with it concepts such as open relationships between equals. The services provided within the network are reciprocated. Here we see one party in the network making judgements about the needs of others in a hierarchical way. Indeed, one image which came across strongly was that of a *family*. T used this image explicitly: 'it's very much a family here, you know, and I wouldn't want to sort of upset anyone, so I might take steps to find out if other people have comments'.

Again later on in the conversation: 'I mean, some of it may be because of that. Don't forget it's only a small organisation, so it is very much like a family. And there are all the sort of intrigues and nice, you know nice things and bad things about families. You know, petty rivalries and all the, you know, I don't have any problems with that, that is actually although they don't sound as, it sounds as though I'm being critical it's not, I mean that sort of thing makes it good. I feel like that anyway.'

There were thus elements of a dual culture, each with similar features. Care and obligation feature in both stereotypes but one is centred on reciprocal behaviour between equals and the other on a more hierarchical relationship. They are broadly compatible but sometimes work against each other. Thus networking, collaborating and acting in a supportive way fit with positive elements of a family culture which create a caring and secure home. At the same time, the controlling and emotionally demanding elements of the same metaphor can produce an over-reliance on some individuals, which they may cultivate, and an unwillingness to adopt new approaches. The paternalism and family image, plus the longevity of employment of some staff, produced a resistance to change which sat uneasily alongside the espoused flexibility and pragmatism. T tried to change a number of clerical practices and found this difficult: 'I think. This is going to sound negative but it isn't really, but there are lots of things that I would like to change and some things I have tried to change that I've really been blocked, not from the management as we call them, just to be funny, but from people below.'

T observed a wish for security, for things not to change, and acknowledged a somewhat schizophrenic view of 'management as we call them, just to be funny'. The two cultures create such a dual view: management may be equals simply carrying out different jobs, or management may be more knowledgeable 'elders'. In a formal sense the management are management but in the culture of the NEC, it appears odd to use that as a general term. The College is a small not-for-profit firm operating within conventional commercial practices, but the cultural style, in some aspects of its work, is more like that of a Cambridge college with responsibilities *in loco parentis*. This style was shifting during the period of the fieldwork.

In the resistance to change, however, there had been some 'hardening of the categories' despite Michael Young's earlier claims. T talks about being 'blocked . . . from people below', indicating how the culture which valued collaboration could potentially stifle activities.

E, the IT specialist, also observed resistance to change at an operational level: 'In the main? Yes. As I say it's better with people who've been here for a short length of time. It's the people who've been here for fifteen, twenty years and they don't see the immediate benefits of change. Having said that, you know, I'm not one for changing systems just for the sake of it, but there certainly are, have been some fairly arcane practices which need to be changed. Yes, at some stage certainly there have been occasions when I think management have actually had to push fairly hard to persuade people to change.'

What we find then, alongside the pragmatism and responsiveness of the organisation, is a resistance to change. This is coupled with a view of the College as a close-knit set of colleagues who have obligations towards each other and reciprocal relationships which to some extent override the formal contractual relationships. Consultancy and contract staff and the little people are treated with care. Relationships inside and outside matter. But there is a tension between the old Cambridge college paternalism and an egalitarian style of networks and partnerships.

One feature of the NEC which reflects the resistance to change just outlined is the departmental structure which is remarkably persistent for such a small organisation. L, the Assistant Director, for example, said a number of times that he was unable to comment with confidence on a particular point because the topic under discussion did not affect his department and he therefore had limited knowledge of it. The departmental boundaries must therefore be relatively strong. All respondents at some time referred to the departmental structure. It was noticeable to E. His job was to begin the implementation of the new information system and he had to work across departments.

This problem was recognised by all respondents in different ways. The College Director was keen to catch the market awareness which fell between departments' responsibilities. T, the tutor-co-ordinator, had very positive comments about attitudes and the company's wish to get things right and she saw departments as internally open. She saw real consultation *within* departments as a normal procedure. But working between departments was difficult. L, for example, even as an Assistant Director, took it for granted that he had very limited knowledge of other areas.

Cultural meanings at the NEC are therefore complex and sometimes ambiguous. It is a genuinely caring institution which slips into paternalism. There is an informal, open atmosphere but a strong sense of department. There are a large number of people associated with the College in various ways who are valued for their intelligence about the commercial and educational environment, and are also valued, using a different sense of the word, as colleagues and people. The treatment of the little people, with the ambiguity of meaning in that phrase, encapsulates the contrasting elements of the culture.

Working in networks and partnerships in general fits comfortably with the culture. Such forms of working depend upon reciprocity and obligation but there are elements of the culture which do not always sit easily with the strategic routine. The tight interpretation of the departmental structure produced a different set of relationships which could hamper the development of networks and, as we shall see, the implementation of the information system began to reveal some of the contradictory features here. Similarly, the family metaphor, which fits more consistently with a departmental structure in which 'family members' get on with their own thing, could be damaging to an open and equal approach to networking.

## Financial imperatives

A number of features of the College's history and environment were implicated in the development and reproduction of the strategic routine. The financial position of the College came up in discussions on many occasions and its history has clearly been heavily influenced by financial imperatives.

Pragmatism and networking have arisen partly from the financial insecurity which has typified the College's history. The external world has traditionally exerted influence through financial pressure. This has tempered pragmatism and a radical agenda with a cautious conservatism. The College Director said: 'But I think NEC always will be a bit behind in that area, because we haven't got a big, you know, amount of money to invest, you know, we're talking when we're thinking of a new course we're thinking "Can we afford to spend twenty thousand?" You know it's not, we're not talking about two hundred thousand or half a million or a million or, you know, it's tiny tiny little budgets that we're having to make go an awful long way and as soon as you move in to anything other than ordinary, you know, you're having to talk about a budget that is ten, twenty, a hundred times. You know, we haven't got that sort of money although, you know, we have had a lot of approaches from other organisations who sort of look along our curriculum material and say "Wow, this has saved us a lot of work", you know, so there's no shortage of potential partnerships. But I think that if we took a very radical or dramatic approach we would lose.'

Current financial success has allowed more time and opportunity for reflection on strategy. A spate of reviews were carried out at the end of 1995 and the beginning of 1996. And each of them was carried out by someone who already had a relationship with the College. They occurred at that time because of the Treasurer's departure. (The reviews were not held up *by* the Treasurer. They were not undertaken until he had left because they would have been difficult for him and others. The College's concern for its staff is exemplified by this.) However, a prime reason for the reviews of the College (rather than their timing) was increased financial security, plus the College's increase in size which prompted a belief that more systematic methods should be employed. The reviews may signal a move to a less opportunistic approach.

The recent financial success had been important and noted by A, the recently recruited Information Technology Manager. 'No, resourcing is not a problem. Because the way that we're structured . . . we are a non-profit-making organisation so any profit, paper profit that we make, has got to be put back into the business, and that's the state that we are in at the moment. We are making a very good surplus and we can't pay shareholders. We've got to put it back, we've got to spend it on something. And, you know, it was expensive to put the link across to the warehouse and another company might not be able to justify that. We actually had the resources to do it, for example. That's just taking one example of quite a costly thing. We could have managed with the old server – another company might have had to manage with the old server – but we have got the surplus to spend on that.' A had worked as a teacher and for commercial organisations before joining the NEC. His perception was that the College was able to resource most of the things it needed – certainly in the IT area – and this was not something he was used to. It was also new for the College. The first comment in this sub-section from the College Director was made in November 1995. A's comments were made in August 1996. In that short period there had been a palpable improvement in feelings of financial security. This came partly from the financial climate but partly from the increasing awareness of the financial situation that had been made possible by staff changes and the use of financial information through the new information system. In general, the resources were being used for investments in, for example, the IT 'link across to the warehouse' or to reduce fees with the intention that all NEC people, including students, should share in the benefits.

Thus the deeply embedded culture and routine practices of using partnerships and networks partly arose from a history of financial insecurity. More recently staff changes and the sharing of information, made possible by the new information system, had enabled staff to manage the new-found financial security more easily. The use made of the new resources has confirmed the College's culture and strengthened its IT base.

## The physical environment

A further feature of the College seemed to play a larger part in routine behaviour than might normally be expected. The main building has a symbolic significance and is the College's major asset. It is implicated in the reproduction of the strategic routine.

In its early life the College inhabited a number of different temporary buildings, moving from one barely acceptable location to another. Its current home, which it owns, is a delightful Victorian building (de Salvo 1993) about 1.5 miles from central Cambridge. Ownership confers security and the attractiveness of the surroundings, much enhanced by the efforts of the current College Director, make the building feel more than simply an office block. It is not a convenient building for its current level of occupancy and the needs of high-tech cabling. It feels a little like a large family home. I questioned O on a number of occasions on why the College did not move to surroundings which would be apparently more convenient and more suited to its current needs.

'Well we've obviously talked about it. I mean and it's come up on several occasions. I think that, I mean I don't feel too worried about the pressure on space. Because what I encourage people to do is to say "OK. Is there another way of getting this done?" I think when people are over-stretched or when ... we hear a particular programme is going to expand or, you know, we've got, you know, a new sort of spurt of enrolments or something like that the first, sort of, knee-jerk response is "We need another member of staff." And the first request is always, I've noticed, for either admin or clerical support. Which is interesting. You know, not, "Is there a program that we can bring in that will actually cope with this routine work, so we can release existing staff to do more interesting work?" It's "We need another person to input." Or, you know, whatever the task that involves, really. And I think it is actually quite a good discipline for NEC to say "OK well where are you going to put that person?"'

O explained that the security provided by owning the building had been a major factor in the continuity of the College during previous financial crises, and she acknowledged that the thought which had gone into the building's decoration went beyond that normally associated with workplaces. To some extent this may simply be one way in which O puts her mark on the organisation. She has undoubted flair in these matters. But it also confirms the way in which the building helps to confer meaning on how the organisation views itself. The physical environment mirrors the culture and routines of the College.

And O agreed that the shape of the organisation was being partly determined by the shape of the building. It would be nice to have more space, she felt, but there was a strong reluctance to consider moving from the building. Expansion into neighbouring houses would be welcomed but was impossible. Ways of working were, therefore, partly determined by the physical environment.

There is a sense of location internally too. L, the Assistant Director, pointed out that shortage of space triggered rethinking about appropriate groups: 'Yes well there was the idea of creating an IT unit rather than devolving IT functions around the building to different departments and that created a need for the space started in one place, and that made it worth while looking at the way in which we grouped. And I'd already been lucky enough to have all my core editorial staff working with me on the same floor and it's made an enormous difference as far as I'm concerned, they can all work together. And we ended up having a similar approach to each of the floors now. [O] and her team are on the first floor, one half of the first floor has been replaced with another half of the first floor. [A different Assistant Director] and his team have the ground floor . . . so it works very neatly . . . which wasn't the case before.'

Movement of staff to different offices around the building had, as in other organisations, created some dissatisfaction among the staff. O tried to ensure that everybody gained something from the move, even if only a redecorated work space. The move was occasioned by the need to accommodate the new IT department but for some staff it seemed to have a more symbolic significance. T saw it as a major disruption beyond the needs of the IT arrangements, and it may be the case that O used the move, non-deliberatively perhaps, to 'unfreeze' people so that other changes might be made more easily, though she never stated this explicitly.

The lack of space in the building would have been a massive constraint if freelance staff were not used. As has already been discussed, freelance arrangements suit the College for other reasons but without a larger building no other option is workable. Decisions about the physical environment thus have repercussions on the organisational structure. The preferred routine of networks and partnerships meant the building could to continue to meet the needs of the College, though not without major challenges for the physical implementation of the information system. The strategic routine allowed the College to continue to occupy a building which was, in some senses, too small. And because it was too small, it confirmed the validity of the strategic routine.

## The implications of technological change

The new information system (IS) has revealed issues about structures and relationships at the NEC and has enabled behaviour to change in ways not hitherto imaginable. For example, the accessibility of financial information has enabled strategy to be more readily evaluated. This sub-section examines the extent to which the IS revealed aspects of the strategic routine or revealed links or contradictions between strategic and operational routines. It also considers the way in which technology was implicated in the changing operational routines and how those were linked, if at all, to strategic practices. A substantial electronic information system covering the whole range of the College's work introduced a new dimension at the NEC. The College was still coming to terms with the implications of the system during the fieldwork and the repercussions, interconnections and changes associated with it were being worked through.

Three elements are particularly noteworthy with respect to the IS and strategic routines. Firstly, as noted already, the IS made it possible to use management information in different ways. Secondly, elements of informating, described below, are apparent through the making explicit of previously tacit knowledge and the use of that knowledge by a wider group. This has challenged some of the taken-for-granted assumptions about appropriate ways of doing things. And thirdly, existing structures have been revealed explicitly, making it possible for questions to be raised about them; for example, the benefit of keeping departments independent of each other has been challenged directly by the use of the IS.

### Informating and IS implementation

The technology cannot be interpreted independently of the people who work with it (Latour 1996a, 1996b). Different views existed about the value of the new information system and the IT department found itself pulled in different directions. Disagreements were not explicit. They arose as grumbles or comments on how things might be. The IT department was implementing what it interpreted as a technical feature but found it almost impossible, at the beginning, to satisfy demands. The demands were not for technology but for service. Technology as artefact and technology-in-use were inextricably mixed.

Initially, grumbles were also substantially increased because of problems in the implementation of the IS. The following examples show how attitudes to the IS changed. The sense staff had of themselves was closely connected to their perception of the system.

In February 1995 the following conversation took place with T, the tutor-co-ordinator:

*Interviewer*:  So in terms of the drudgery, as you might to say, it's the removal of drudgery.

*T*:  Except there isn't, you see. I was just going to go on to say that there isn't a removal of drudgery.

*Interviewer*:  There should be.

*T*:  Yes.

*Interviewer*:  But it hasn't happened?

*T*:  No. And I just don't know. Because the IBM, which was the system I came into here, is so old-fashioned that it's almost laughable. But it did its job. But it couldn't report. You know, it was pathetic really. You know, I mean it was really years out of date. . . . But the whole thing is written round the accounts package. And that's not necessarily what the other departments want. So everything else is almost a refinement of the system, and you have to wait your turn to get it. And it's really pathetic the way that they've sort of transferred information in the same format as it was on the IBM, and in actual fact that was the way, worst way to present information. I just would have gone for a clean sweep if it had been me.

So attempts to maintain continuity between systems, in a sense to mimic the familiar (in the way that was successful for Digital when it introduced tele-working), were not always welcomed, particularly when they slowed down activity. T, who was relatively new at the NEC, would have 'gone for a clean sweep'. Developing an old accounts package to make things better was 'really pathetic'. By November 1995, the problems had become potentially very damaging:

*T*:  That's why I'm in the state I am today, you know, because everything is a battle. I should be writing a report. . . . It's a real problem for me.

*Interviewer*:  This is almost entirely because of the IT changes?

*T*:  Well yes because I used to be able to [description of an administrative system]. Well because of something else not working on the system I can't tell who are inactive and who are working students. It's unbelievable. And so they've not had follow-up letters, they've not been communicated with, we can't call groups of students up to mail them because we don't know who's working and who isn't. I just . . .

Systems problems were thus making life extremely difficult for T and she was angry. Her ability to do things had lessened as a result of the current state of the IS and she found this discouraging and, I sensed, a blow to her pride in her ability to do a good job without having to ask for support and help. She continued: 'And there is two schools of thought, I mean thought within the organisation about this. There's one school . . . who think that everything should be controlled by the IT department and nobody should be producing things individually. And there's another school who thinks that providing that there's an agreed time for printing out huge reports and huge label runs, which I think we all accept, which may have to be done at the weekend or overnight, then the control should be within the individual, because of course it's taken away a lot of things that should be done by the individual and of course being motivated and you know there's some loss of tasks there for the individual.'

There was a lack of control for T. Jobs she had carried out before, which had been under her control, were potentially being shifted elsewhere. This reduced her motivation. I interpreted her expressions and manner as a feeling that she was being demeaned. She was always bubbly and lively, and sometimes quite blunt, but on this occasion there was underlying tension, uncertainty and a sense that she was being treated badly by the organisation. She was becoming, in some respects, an operator of the machine, waiting for it, or for others in the organisation, to deliver, when she had been used to doing all these things for herself. This is one feature of informating (Zuboff 1988). T felt threatened by a form of deskilling which took away aspects of work which had previously called on her intelligence. There were other features of the shift of control too. 'And also because of the lack of control it, you know, it can create insecurity, can't it? I mean actually in the day-to-day work because you don't feel in control. Can be a problem. It's certainly a problem for me 'cos I never know what the damn thing's going to do next. That's how I feel about it. I mean I'm the first one, I mean I, there's no protection passwords, there's nothing. Everybody can get into every bit of the system which is terrifying. And they do. Because they get in accidentally. I've been all over. And I don't want to go all over. You know, I just want to go into the bits that I need to go into.'

Other features of informating are implicated here. 'Everybody can get into every bit of the system', so those things which previously resided in T's head or on her own (un-networked) database were now available to everyone. 'There's no protection passwords'. This was 'terrifying' and although the reason for the terror was unspecified, it was clear from T's manner that she feared both that other people might mess up her data and that she might inadvertently create problems for others. In addition, she was concerned about what precisely she could now regard as hers. These are also potentially empowering by-products of informating but for T, at that time, they were not viewed positively. However, they reveal the way in which the new IS was beginning to challenge existing practices and the structural relationships individuals and departments had with each other.

The example shows how the autonomy of departments was threatened by centralising IT in a new department. Old assumptions of autonomy and control created conflicts. Some of the problems arose from poor implementation and training but they also arose from the changed nature of the system. The informating potential and loss of control were major problems for T.

But the information system was intended to make information more widely available and to facilitate some of the changes which T found challenging. E, the IT specialist, described his concept of the importance of the new systems to the NEC.

E:             I think basically they're very important because what we tend to do here is mainly administrative work and a lot of it is fairly dreary and routine, and I think electronic systems obviously cut down a lot of the day-to-day drudgery. Free people off to actually talk, communicate with students and tutors. Which is something we're not very good at doing, I think, partly because people's time is spent doing these more mundane tasks. So I think there is a lot of scope and there has been a lot of scope.

*Interviewer*:   Cut out the drudgery. Is that, do they do more than that?

E:             They will do I think. I mean again that was something that was always an intention that we should do far more analysis work, provide much more information about our students. Do proper forecasting, budgeting, et cetera. Which is something that the old systems never did. The new systems will be much more capable of doing that. . . . And also at their desk. The old system there was, sort of, one terminal per room, so you were obviously restricted on who has access to the terminal, who has to actually use the machine. Whereas now people have actually got their desk top. The idea is that they can import data . . . and drop it into any application.

The IS was thus intended to provide access to information for a wide range of staff. But increasing access to information, and making information clearer, challenges existing structures in ways which had not been anticipated. At the time of T's comments and E's explanation (which preceded T's experience by six months), the implementation process had not been completed and the system in place was slow. Subsequent upgrades and user support resolved much of T's concern. She also absorbed the changes and began to work in different ways. Six months later she said: 'Well I think it was, to talk about the IT. I feel much happier about it. There was quite a lot of meetings. . . . And I feel more relaxed about it. Sort of can see the end. I still could be, I mean I still could be critical about it, but there's no point in being, it's better just to hope that it's going to get better and better.'

E's enhancements had solved a large number of detailed problems which had been very irritating to T, and her continued use of the system had reas-sured her that her initial terror was unjustified. But the enhancements had changed the system too. E's ability to adjust the technology-in-use and T's increasing confidence meant that both T and the technology had changed through the interaction between them. The IS was a tool moulded by human endeavour but it also possessed some features of agency. The technological tool did not possess intentionality. It could not decide to carry out particular functions but its particular form had caused T to modify her behaviour in ways in which she would not have done without that specific IS. This is not a determinist argument. The fact of the interaction is illustrative of the way in which both human agents and technology are implicated in social change (Latour 1996b; Jones 1998b).

The IS also focused attention on structural features which had been taken for granted before. The rigidity of the departmental structure was an issue which came to the fore in thinking through the IS. Implicitly, the IS challenged previously held notions of autonomy and responsibility. T's responses are one example. The College Director had reflected on such issues and early in the implementation phase made it clear that she was looking for a central IT unit. The then current spread of IT functions in departments was a problem, as she saw it, because the opportunities to gather information received by one department but only of value to another were not being taken. Information of high value to the marketing department may be of no interest to the finance department, but it arrived on invoices. This created a potential for conflict between the two departments with one insisting that it did not have the resources to code all the details it received and the other arguing that such information was vital. Centralising IT would enable such details to be captured, and in doing so the skills and knowledge of the different departments would be shared. Information previously held in the memories and filing cabinets of particular staff or groups of staff would be available via the IS. The interpretation and use of that information might change. Informating was thus one objective of the IS, though it was not expressed in that way, and was seen as a pragmatic response to potential changes. O was anticipating a more flexible structure which would break down departmental barriers. This would bring the internal structure more in line with her perception of the College's external environment. L also saw the new IS impacting on the departmental structure.

'I think on the whole change is really straightforward without any sense of shock, at the move from entirely paper-based communication to e-mail, for example, has been extremely smooth and effective. I think a lot of it is finding it difficult to let go of the idea of keeping paper copies of drafts and redrafts and so forth. So we did have a literal sort of spring clean at the beginning of this year when we got rid of a lot of. . . . Committed ourselves to keeping things on disk rather than on paper, and perhaps that was a culture change for some of the people here.'

He continued: 'And perhaps move towards a looser management system whereby there are people from different departments [inaudible phrase]. We haven't got structures in place that make it informal easily. It does happen in fact already in many ways, but we don't have the system.'

## Was the IS a causal factor?

L, the Assistant Director, saw benefits from the potential of the IS for sharing information – a feature of informating as privately held information enters the public domain – and breaking down boundaries. He was inexperienced in IT matters but had become enthusiastic about them and had enrolled on a number of high-level courses dealing with new approaches to communications technology. Some things were already happening and he hoped that easier communication and the sharing of information across the institution might become more widespread and a looser structure might come into place. L and O, at their more senior level in the institution, were thus trying to create the changes which (unintentionally) caused many of the problems experienced by T. L saw the IS as one of the *causal* factors in these changes.

The College Director argued, however, that the potential of information technology was not driving the changes:

*Interviewer*:   How much of the changes were driven by the opportunity for a new IT structure?

O:   Probably that was not the deciding factor. I mean, you know, you could look at the changes in two ways. I mean one is [the Treasurer's] departure being the catalyst, which it was. . . . But I mean certainly the IT system has had an impact, but it wouldn't be honest of me to say that that, you know, it was driven by the IT system. And I think we do see the IT system as being a tool, you know, rather than something that we have to, sort of, organise our business around.

The information system may have been seen as a tool but it was clear that trying to maintain old operational routines under the new system would not always work. The new Finance Director was reported by O to have found old and new routines getting in each other's way: 'I think I mean one of the things that [the Finance Director] said to me about the accounts package, and he's come in, you know, fairly new, is that he can see now, having been in NEC for, he started in August so, you know, sort of, a bit over three months, that some of the problems and some of the things that the accounts staff complained about with the new Tetra system was because they were trying to do things in the way they always had done. And it was a tremendously complicated feat to make the Tetra software perform in that way. And so, you know, they were finding the system unnecessarily complex.'

Attempts to stay with well-tried operational routines were thus challenged by the new system. As staff learned about the system, and it was adapted to their needs (as exemplified by T's experience), the complexity diminished.

O, however, was missing some of the shifts which were occurring when she argued that it was only 'a tool'. The interaction of the IS with existing routines was changing the organisation of the business. It did this for individual staff, such as T, and it also raised questions, perceived only implicitly before, about the ways in which people carried out their work. In a conversation with E, then an IT consultant, and A, who had just been appointed IT manager, it became clear that their perception of their role was as providers of a new tool, but that in fulfilling it awkward questions sometimes came up. E commented: 'It's interesting that having computerised some areas that weren't previously computerised and [it] has actually identified these problems.' (The problems were of many sorts, for example that data could not be incorporated in certain ways, or that particular individuals did not know about data which had implications for their work.) A made a similar comment: 'That's right, but we're doing something which is perhaps beyond our brief here, we're identifying inefficient systems within the company.' He continued a moment later, 'That's definitely true. Yes. I mean I've noticed that since I've been here. I've had to not just solve problems, existing problems, but think of new ways of doing things and if the IT unit couldn't do them then they wouldn't happen.'

A did not specify an example at this stage in the conversation. He was surprised that he had become entangled in what were, in some cases, issues of policy. It is clear that the IS was not a neutral factor in changing routine behaviour, however. From the interviews, staff believed that its existence made it possible for things to happen which had been impossible previously. Its existence also raised questions which had not previously seemed significant or of which people had been unaware. While the IS itself was not driving change it was clear that both recursively and discursively change was taking place which would not have happened without the IS. Such changes did not relate only to technological solutions. The technology was implicated in changes in social systems.

The changes taking place were directly affecting operational routines and questioning existing practices. The internal culture was shifting slowly and some challenging questions were beginning to arise. For example, one specific feature which was mentioned towards the end of the IS implementation period was the fact that O did not use a computer. This fact had had no relevance earlier and its implications both practically and symbolically for the College had simply not existed. But now that the IS was beginning to play a bigger role, such matters became noteworthy.

The fact that the Director did not use a computer and was not therefore able to stay in close touch with what was becoming a major communications and information network in the College raised questions about the nature of communication and the perception at a senior level of the value of IT for strategic relationships internally and for building networks and partnerships outside. It could potentially be an important symbol. It also raised questions about the direct experience she could call upon in assessing the implications of the IS. I raised this with O. I asked her if the lack of connection to the IS didn't make her feel uncomfortable and if perhaps she was signalling something about her preparedness to be part of the organisational culture. She replied:

'I don't feel uncomfortable. I mean, you know, I have decided that I will and that, you know, I'll do so early next year. But I mean, I suppose it may reflect a feeling that it can become very, very, sort of, time wasting . . . And I mean of the things that I am terribly short of is time. You know, and I have to struggle quite hard to, you know, sort of cut away the inessentials and just to concentrate on the essential things. And, you know, I suppose the thought of coming in on a Monday morning and finding there's thirty or forty e-mail messages, I don't know. You know. I suppose I think if somebody really wants to communicate with me they'll put their head round the door or they'll telephone . . . And I don't particularly want to get involved with the banter. Maybe I'm arrogant, I don't know, you know. I mean, I suppose I don't want to sit and have to respond to thirty e-mail messages before I even start the day.'

The decision to acquire a modern electronic information system had thus begun to influence, or at least question, the routine behaviour of the College's senior manager. The College had changed not just because work done previously could now be carried out in a different way but because the nature of work and relationships was changing through the possibilities opened up by the IS. Modifications in social systems then created new legitimate uses for the IS, thus modifying the possibilities for the technology (in use). If O did not develop her IT skills this would give messages to the College about what was and was not appropriate. If she did develop them, similarly, the appropriateness of different kinds of behaviour would be redefined and different possibilities would be created.

In discussions with other members of staff, O's lack of computing skills did not appear to be very significant, nor is she backward in considering the possibilities of the new technologies. She did not, however, envisage that the institution of a much bigger IT function would have a profound effect on the way the organisation operates: 'The IT unit is a unit that services, in the same way that the accounts department does, you know. It provides a service to the organisation, print comes into the same category, and it doesn't, on its own, drive the organisation. But at the same time from a management point of view the, you know, there is a need to make sure that we're using IT in the most strategic way. But I mean I still feel that that is to support the work of the organisation. Because, you know, we're not an IT company.'

It may be that O will develop IT skills and become involved with changes that transform the NEC. As a key agent in the College, her attitude towards IT practically and conceptually will be influential. During the fieldwork there was no evidence that O hindered the implementation of an IS strategy; however, it was only towards the close of the fieldwork that her attitude to the use of computing became worthy of comment.

Her comment above, however, does not accurately reflect what was happening in the College. The IT department is not the same as the accounts department. The IS enabled different forms of communication to take place, and information to be shared in different ways which challenged existing taken-for-granted assumptions, and raised questions which had previously been implicit or unrecognised. An accounts system can do some of these things. It can certainly raise questions and challenge assumptions but the accounts system is a report about relationships, it is not part of those relationships. In contrast, the IS meant that human agency and the (computing) network changed through interaction with each other.

The information system was not causal therefore in a technologically determinist sense. But as a factor in changing operational routines such as those carried out by T, and in challenging the internal structure, it has been implicated in changes which have potentially wider dimensions. The College Director's initially limited connection to the system has been challenged and this could presage major changes in the way she operates internally and externally. The decision to purchase, develop and implement an information system was itself strategic insofar as it was intended to enable the College to interact more effectively with its students and clients by, for example, picking up key information, but the IS was seen as a means to allow information to flow and be shared better internally and to improve co-ordination of primarily operational functions. Those changes were very nearly in place when the fieldwork ended and what seemed clear was that developments would not finish there. The shifts in structures and practices were likely to bring changes to the College at all levels as each acted back on the other. At that stage the new information system had not been *directly* implicated in the strategic networks and partnerships routine but, in making some features of the College explicit, it enabled challenges to be made about appropriate behaviour, some of which had potentially strategic significance.

## Networking again

Networking had different meanings at different times and in different places in the NEC. In interpretations of the interviews, at least three uses can be found, as the quotations given throughout illustrate. A common use of the term referred to the network of computers within the College, with occasional references to outside computer networks. A second usage was by senior staff to refer to the range of different kinds of staff who work with the College. Partnership was also used in this context. The third usage was primarily by the Director and referred to the broad range of contacts and 'grapevines' to which the College was linked in different ways. Partnership was also used here to refer to some of these links. Alongside the latter two usages, the notions of obligation and reciprocity were also implied and sometimes explicit. The term was not conventionally used for internal staff relations. There the departmental structure and family or paternalistic culture appeared more appropriate. The staff were part of a wider network and sometimes formed partnerships externally but did not appear to see themselves as a small internal network.

In my initial analysis I assumed the first usage – a computer network – could not be useful in assessing the College's strategic routines since it is a specialised computing term but the fact of that network and its potential for compatibility with the strategic routine provide valuable contrasts. In practice the availability of the computing network and its associated opportunities for sharing information constituted an important factor in revealing some of the tensions between the desire for an open internal network, espoused by the College Director and Assistant Director in particular, and the difficulties this created for more junior staff. The network of computers fits well with the strategic routine but in doing so raises explicitly some of the differences between that way of working and the 'family' or paternalistic model. We saw evidence of stress and tension created by the sharing of information via the network and the granting of access to information previously held by others. The network and the agents involved then interacted and the tensions were reduced. At that time the computer network external to the College was largely unexplored.

Thus the informating potential of the IS at the NEC was implicated in changing operational routines, and some interactions with the IS uncovered cultural conflicts which showed that the strategic routine based in partnerships and networks did not operate in the same way throughout the organisation. The contrast between network and family is an interesting one. Network implies equality and the use of a computer network can also carry this implication. Family can imply control and computer networks can also be used to control. The departmental structure comes from a paternalistic, *in loco parentis*, Cambridge college culture and was resistant to some of the implications of networking. The second two usages of network set out earlier are part of the strategic routine as identified here but they are not entirely compatible with the family metaphor.

The term 'family' was volunteered only by T. Other participants broadly approved of it when I offered it to them but I sensed a more ready acceptance of it from less senior staff. In their actions, however, there seemed to be some ambivalence. O reflected in various ways on her care for the building and agreed that it was in some ways an extension of her home, though kept very much separate.

None of these things are significant in themselves but, taken together, show that the culture which has been characterised here as a family or paternalistic one, in which big people look out for little people, retains some influence in the College. Staff appear to operate in both cultures in different proportions and with different emphases.

The use of an information system revealed some of the tensions involved in the different metaphors. The IS is implicated in changes at different levels but the changes can only be effectively analysed by looking at the information and social systems together.

# Summary

The NEC case shows that routine behaviour exists at strategic levels. Not all strategic behaviour is routine but the strategic practices adopted at the College confined its range of strategic behaviour to a relatively narrow band of options. In Nelson and Winter's (1982) terms the networking strategic routine was involved in search primarily in the external environment.

Strategy was defined in the introduction to this chapter as relating to the organisation's relationship with its external environment, specifically the organisation's attempts to shape its internal and external environments so that it is better placed to achieve its objectives or to redefine its objectives. The institution of the new information system at the NEC was strategic in that sense. It was adopted in an attempt to shape the internal environment so that it (the internal environment) was better placed to meet the College's objectives. There is thus a link between the operational changes which were associated with the IS and the strategy of the College.

In tracing through the strategic routine here and the practices which were related to the IS, a tension can be seen between the contrasting meanings at the NEC of networks and associated egalitarian relationships, and family and some relatively hierarchical relationships. All respondents appeared to adopt aspects of each meaning at some time but there was an espoused desire from senior managers to see the networking culture adopted more thoroughly in internal relationships, a sense in which they wished to bring external and internal practices into alignment, though they did not set it out in this way. Partly by design, but more extensively through the practices associated with the IS implementation, the opportunities for networking internally increased and the structural, cultural and routine blockages to internal networks were increasingly challenged. The connection between operational and strategic routines is thus illustrated.

The College has been described as opportunistic, pragmatic, flexible and built upon obligation and reciprocity, and also, though still demanding obligation and reciprocity, traditional, paternalistic, hierarchical and showing some resistance to change. These practices which arose from cultural meanings, the College's history and financial imperatives were developed by strong individuals. They were reproduced partly through the constraints of the College's building which itself was symbolic of the meanings through which College members defined themselves and their activities. Technology was partly constitutive of the practices adopted. It did not imbue every aspect of the College's work as at Digital, but despite the claims to the contrary, the challenges to taken-for-granted practices associated with the new information system did change the environment in some unintended ways. The organisation's key competences of providing distance-learning courses in student-centred, accessible ways are not linked to technology directly, though they do appear to be enhanced by technology. There was a tension which was fought out over implicit taken-for-granted ways of working rather than explicitly. Both T's and O's relationships with the technology are exemplary in this respect. Judgements which define the technology solely as a tool do not take into account the ways in which the technologies and the human agents interact with each other.

There are thus particular established, significant and sanctioned practices within the organisation which predispose it to act in certain ways strategically, namely to seek out partnerships and use carefully developed networks. The organisation *tends* to act in that way. The practices are not absolute, unbending patterns. The networking and collaborative style which is celebrated by the culture feeds through the organisation's structures and the actions of its agents into behaviour which is identifiably the 'NEC way'. In order to understand strategic behaviour it is necessary to understand this NEC way of doing things. As with Digital, the recursive relationship between structures and agents confirms the practices and modifies them. The practices are subjected to discursive assessments but only within a sense of the appropriate ways to act and react.

# 6 Unipalm-Pipex: change as routine

## Introduction

*Interviewer*: Well OK, what hasn't changed? What's the same, or what's identifiably Unipalm that was here two three years ago and is still here?

*P*: To give you a rather devious and nasty answer, what hasn't changed is our ability to change everything at the turn of a hat, at the drop of a hat.

The answer to my question given by P, a senior manager at Unipalm-Pipex, encapsulates the view of the company of most, if not all, of the people connected with it. Unipalm-Pipex was an extraordinary company. It was involved in the formative development of a new industry. The changes it underwent were large, and sometimes appeared dramatic. If P is right and the company had an ability 'to change everything . . . at the drop of a hat', can this be considered a routine which the company had acquired or learned or was the company simply forced to change because of the turbulent environment it faced? How did the company acquire the competences to change everything, if it did, and how was technology implicated in this change?

The concept of routine as an established and sanctioned practice is not normally associated with change. There is a sense in conventional usage that routine and change are opposites. However, the argument here will be that change and routine are compatible and that the explanatory value of routine is enhanced by expanding its meaning to incorporate change. The concept of routine will be stretched – tested almost to destruction – in evaluating whether Unipalm-Pipex should be viewed as being tossed about in an environment of significant change or as an organisation which developed ways of incorporating change into its taken-for-granted behaviour.

The account here takes the distinction between operational and strategic routines as read and develops the concept of strategic routine to examine whether taken-for-granted practices at a strategic level can take into account rapid change. Is it possible for companies to develop, or acquire, strategic routines which enable them to cope with and, potentially, benefit from change, or are routines, by their nature, unhelpful in novel circumstances? As before, there will be a focus on the way in which technology is implicated in the explanation.

*Background to the company*

Unipalm was established in 1986 as a software distribution company. It was based upon an approach to networked computing which argued that systems should be open, that is essentially accessible to different kinds of computer. The company acquired distribution rights to system software which permitted openness. Specifically, Unipalm became a distributor for TCP/IP (Transmission Control Protocol/Internet Protocol) which is a means of transporting computer data over virtually any medium. D, the company founder, and his partner acquired the distribution franchise for FTP Software Inc., an American software company which produced TCP/IP software. This was to prove a highly profitable acquisition. In 1986 computer networking was developing but was small scale in comparison with the huge growth which has occurred subsequently.

During the early 1990s the demand for network software became enormous. Unipalm's link with FTP gave it the ability to provide the software for which everyone seemed to be asking. It was able to generate cash and to support other developments on the basis of the FTP franchise. In the mid-1990s network software became much more readily available. Microsoft's Windows 95, for example, gave it away apparently free of charge as part of the bundle of services included in the basic product, but by then Unipalm had changed its product strategy and was no longer dependent on distributing TCP/IP.

As well as TCP/IP Unipalm provided other software products primarily linked in a broad way to computer networking. At the time of its formation, the company deliberately developed a series of connected businesses, with different brand names, which were planned eventually to spin off and become independent. The businesses were all linked to the basic Unipalm distribution business but had separate identities. The *Computer College* was set up as a training company, specialising in Unipalm's products; *XTech* was responsible for developing and distributing electronic mail systems; and *Unipalm Consultants* provided technical support and consultancy for customers. The customer base was corporate. At that time Unipalm did not aim to meet the needs of single computer users or the domestic market.

In 1991 it became clear that the five partners then running the company had different ideas about the way the company should develop. D, in particular, was unsettled. He left the company, with no resources at that stage, and began to develop ideas for providing internet connections. At this time the company had around thirty employees. The other directors made no financial offer to him. D had a significant investment in Unipalm and was finding it very difficult to obtain the support financially or commercially which he needed to develop his new ideas. The lack of an offer from his previous partners meant he could not financially realise his investment. After much internal squabbling between the directors, D and his partner bought out the other Unipalm directors using funding from 3i and their joint 51 per cent ownership of Unipalm. The partner took on the responsibility for directing the existing Unipalm distribution company and D used the base provided by the distribution business to develop a new internet connection business within the Unipalm group. The internet connection business was called *Pipex* (Public Internet Protocol EXchange) Ltd.

The distribution business continued to be very profitable and provided the resources necessary to develop Pipex. Pipex was not profitable but was growing extraordinarily fast, after a slow start. The internet had become fashionable and available to a general market, from its original restricted links as a network for the American military and as an international network for academic institutions. D's aim was to become leader of the new market just appearing. In March 1994 the Unipalm group was floated on the stock exchange. Although the Unipalm distribution business was still the primary source of funding, the presentations to city financiers were made entirely on the basis of the growth potential of internet connections.

A shift was occurring in company philosophy, though the desirability of particular outcomes was a source of considerable argument within the company. D's view was that it was no longer sensible to split the company up into separate businesses and a task force was set up to consider the proposal. By the end of 1994 it was seen by some that a single company focused around products related to the internet was an appropriate direction and a primary task became the merging of what were then seen as the two major sides of the company: Unipalm (including the other related businesses) and Pipex. While Unipalm was the primary income earner at this stage, Pipex was more aggressive and saw itself as the future of the company. In July 1995 the company changed its trading name to Unipalm-Pipex.

The merging of the two teams involved different cultures, structures, routines and agents. The merger took place in a turbulent environment in which new products were developing rapidly and in which consumer preferences were immature. Competition was international. The environment was dependent upon policies of major companies such as BT and decisions about telecommunications strategies being taken by national and international governments. A number of large and small companies were supplying related products but there was no easily identifiable market place (Cronin 1994).

Unipalm-Pipex was a profitable company. The Unipalm side of the company was in many respects a cash cow supporting the growth of Pipex. Pipex itself had not made profits, though its share price had risen dramatically, but it was claimed that large profits would be generated once the company's rate of growth slowed and it was able to realise a return from its investment in infrastructure.

The company's *Annual Report* in 1995 described the relative positions:

> Following a decision taken in 1993, the Group floated on the London
> Stock Exchange in 1994. The purpose of this flotation was to raise
> additional funds to expand the operations of The Public I.P. Exchange
> Limited (Pipex), an Internet Services Provider. During the financial year
> under review the Group traded as two principal operations, Unipalm
> Limited (Unipalm) whose activities include the sale of networking
> software and related services, and Pipex which provided Internet services.
> The Group achieved record turnover of £17.687 million up 64%. Profit
> before tax rose 63% to £442,000. . . . Unipalm's turnover rose 49% to
> £13.928 million with pre tax profits rising 214% to £1.530 million. Its
> products include a range of networking software products based on the
> TCP/IP protocol together with training, consulting and internal security
> 'firewall' products. . . . Pipex's turnover continued to grow at a rate of
> approximately 10% a month rising 263% to £3.759 million. Losses in
> the period rose to £1.088 million. Losses are being incurred because of
> the implementation of the strategy outlined at the time of flotation to
> grow the business as rapidly as possible. . . . Whilst the rate of growth
> of new customers is increasing, Pipex is loss making because of
> expenditure on infrastructure (people, equipment and network capacity)
> ahead of customer orders. At any point in time, the removal from
> the cost base of excess capacity, sales and marketing personnel and
> installation personnel would result in a profitable subscription based
> business.
>
> (Unipalm Group 1995)

In the autumn of 1995 Unipalm-Pipex merged with an American internet
connection company, UUnet. UUnet had a profile in the United States which
was similar to that of Unipalm-Pipex in the UK. Both companies were looking
for an international presence. Unipalm-Pipex became the European division
of UUnet (and changed its name to UUnet-Pipex). UUnet Technologies Inc.,
together with Unipalm and its related entities, became one of the world's
leading providers of a range of internet access options, applications, security
products and consulting services.

In April 1996 UUnet announced a merger agreement with a major US
telecommunications provider MFS. The merger created one of the world's
largest business communications companies, providing a single source for
a full range of internet, voice, data and video services over an advanced
international fibre network. The combined company was then the only
internet service provider to own or control fibre optic local loop, inter-city
and undersea facilities in the USA, UK, France and Germany. The ownership
(by MFS) of the fibre optic cabling was a major feature of the merger. All
other internet service providers at that time bought 'band-width' from
telecommunications companies, for example over 40 per cent of UUnet's

Table 6.1  Unipalm Group PLC, company financial profile during the period of the fieldwork

| | 12/96 | 12/95 (8 months) | 04/95 | 04/94 | 04/93 | 5-year average |
|---|---|---|---|---|---|---|
| Turnover | 31,853 | 14,085 | 17,687 | 10,753 | 8,390 | 17,479 |
| Profit before tax | −2,141 | −8,414 | 442 | 272 | 605 | −1,976 |
| Net tangible assets | −4,223 | 1,551 | 7,173 | 5,956 | 1,256 | 2,342 |
| Shareholder funds | −5,737 | −1,399 | 6,541 | 6,179 | 970 | 1,310 |
| Profit margin (%) | −6.72 | −59.74 | 2.50 | 2.53 | 7.21 | −10.84 |
| % return on shareholder funds | 37.32 | 902.14 | 6.76 | 4.40 | 62.37 | 202.6 |
| % return on capital employed | 51.43 | −813.73 | 6.06 | 4.16 | 48.17 | −140.79 |
| Number of employees | 333 | 223 | 159 | 97 | 97 | 182 |

Source: FAME Financial Analysis Made Easy Database.

Note
All figures are given in £000 except where stated; columns refer to the financial year ending in the month indicated.

network expenses were for local communications services (UUnet/MFS press release, April 1996). Significant savings were claimed from the merger, and at the same time the combined marketing teams of UUnet and MFS were able to sell their interrelated product range, thus taking advantage of economies of scope. Furthermore, UUnet (and initially, Unipalm-Pipex in Europe) had a strategic relationship with Microsoft for the development, operation and maintenance of a large-scale, high-speed, dial-up network, which was the primary internet network infrastructure for Microsoft (including the Microsoft Network which is part of Windows 95). Microsoft had a 13 per cent equity position in UUnet and a Microsoft representative sat on UUnet's Board of Directors.

With UUnet's strong position as an Internet Service Provider and MFS's international high-bandwidth network platform, plus the support of Microsoft, the combined company was well positioned to benefit from the shift to internet-based communications. This is the world of *big* business and is a long way from the original 1986 development of Unipalm in a small lock-up building in a village in rural Cambridgeshire.

In the spring of 1996 D resigned as Director, European Operations UUnet-Pipex, retaining personal assets of £38 million. He planned to follow up interests in politics and invested in a consortium bidding for a local radio franchise. (Subsequent to the completion of the fieldwork, UUnet became a subsidiary of WorldCom, the world's largest internet service provider. It operated in over fifty countries and in the autumn of 1997 was the company which stopped BT achieving its world ambitions through its take-over of the US telecoms company MCI (see, for example, Barrie *et al.* 1997). There is a nice irony here. D had often professed the opinion that Unipalm-Pipex could potentially be wiped out by BT but that BT had been unable to get its act together.)

### Staff directly participating in the research at Unipalm-Pipex

Five Unipalm-Pipex staff participated in the research:

D:   one of the founders of Unipalm and ultimately Managing Director of Unipalm-Pipex plc;

H:   a Cambridge engineering graduate who had been involved in network developments as a teacher and consultant for over ten years; H originally moved to Unipalm to help set up a subsidiary but moved into what was effectively a personal assistant role to D. He then took part in the public presentation and public relations activities of the company. He resigned from Unipalm halfway through the fieldwork but continued to take part as a knowledgeable 'outsider';

P:   joined Unipalm in 1993 as a senior technical manager. He came from a big company technology background, one of few people at Unipalm to have such experience. At the end of the fieldwork he was the only participant still with the company and had become Head of a major section of UUnet-Pipex;

W: had worked for Unipalm for five years when the fieldwork began. He was responsible, through a team leader, to the Head of Sales for major corporate accounts. He left the company towards the end of the fieldwork, partly because of dissatisfaction with changes at Unipalm, to work for a company offering similar products;

K: joined Unipalm in 1991. It was her first job. She worked in purchasing and Unipalm sales, ultimately becoming a sales executive for the combined Unipalm-Pipex company. She moved to a new organisation midway through the fieldwork.

## Change as routine

Some of the change at Unipalm-Pipex was directly its own creation, for example change arising from the merger of Unipalm and Pipex. Some arose from its own behaviour in the external environment and an extended example of this, the company's move into the individual dial-up market, will be developed in this section. Some arose from the turbulence of the environment in which it operated.

The company culture was aggressive and fast moving. How did the company cope with change? Change was ubiquitous and therefore routine in the sense of being a normal feature of the company's world. But was the company able to develop its own routines to cope with the manifestations of change apparent everywhere? Clearly Unipalm and Pipex were different. While Unipalm faced frequent changes in software specifications and a rapidly growing market, its environment possessed easily recognised regularities. The distribution business had created established, significant and sanctioned practices using formal and informally developed relationships as well as technologically driven processes via its databases. Many of the day-to-day operations at Unipalm were straightforward. Relatively well-developed scripts had been created for the approach both to customers and suppliers and to internal operations. There was a sales 'bible' of frequently asked questions and tips and clues about products, as well as a database (which changed during the period under study) which was a crucial site of operational routines, on which business was followed through. The distribution business had been awarded BS5750 for its procedures in dealing with customers. There was a consistency in the way employees were allowed autonomy in developing their relationships with customers or suppliers. The payments system (with bonuses based on targets and margins earned for sales) was a 'controlling' device and the database defined many operational routines (interview with H 1995). The world of Unipalm had an operational regularity about it.

Pipex, on the other hand, still saw itself as on a mission, driving hard with few holds barred.

## Technological interconnections and operational routines

Without the new network technologies there would have been no Unipalm. The exploitation of those technologies was its *raison d'être*. D, the company founder, was driven by a desire to be an industry leader. But a focus on technology also drove many day-to-day activities, so that staff, while apparently autonomous, had relatively little room for manoeuvre. Informating (Zuboff 1988) was widely apparent at operational levels. Unipalm had used relatively sophisticated databases for most of its history. These databases enabled staff to manage their work in an information-rich environment. The database provided them with an enormous amount of detail about transactions they and others were undertaking. The information was public to the company and constrained the manner in which activities took place. Autonomy was thus severely circumscribed. The precise manner in which an activity took place was under the control of the staff member but the basic task was highly routinised. W, from corporate sales, said: 'But over the last three years we have migrated away from that into electronic systems. For example, when I joined Unipalm five years ago the bought leads were done by paper. You'd have a stack of leads on your desk, you'd work through them one at a time, whereas now it's all on the computer screen. You click on the next screen, it brings up the information, you call them, you put the new information in, click on the next one. . . . So that's all on-line.'

The database was a commercial piece of software, called Unitrac, which had been bought in to the company. K, the junior sales executive, found it 'brilliant': 'Yes, yes. The most important thing for me as internal sales person is Unitrac, which is a database of the customers which you are probably, it's every customer that sales have ever approached or has ever approached Unipalm. And I mean it must be tremendous now. And you can go in there and you can amend details on it, so if this customer rings you back next week you could click on their name, and all the information that you've spoken to them about, which you've actually put in there, is there. So for a sales person that is a brilliant system.'

Not all routines were directly linked to the software but the approach adopted was heavily influenced by the tracking systems possible in the software. Eventually help pages for staff were built into the system itself though initially they were kept more conventionally. Procedures were not laid down in detail but the electronic system effectively controlled the options and, on the sales side, standard customer relations procedures were in place.

Unitrac was not the first database Unipalm had used. Previous systems had been subject to problems and senior management had experienced difficulty in getting this kind of system accepted. The difficulty arose, according to the staff interviewed, from the unreliability or slowness of the system, not the principle of using such an approach. The introduction of Unitrac had not been free of bugs but its facilities were much appreciated as the quotations above indicate. The use of such a system effectively gave Unipalm a very clear set of operational routines. They were not written down in a conventional way. The routine was built into the software in the sense that the screens and information available in the software acted as an agenda and resource base for staff. Unipalm's memory of what had been done and how its activities were carried out, at a practical level, were also, in that sense, embodied in the software.

Informating creates the opportunity for control structures via the transparency of the activities at operational levels. That is to say, it is possible for everyone to see what is going on. In general, control was not explicitly exercised at Unipalm. It was unnecessary to create formal control structures because the electronic system revealed enough without them. Sales targets and bonuses were tightly structured but the potential for control using the technological equivalent of Bentham's Panopticon (Zuboff 1988) was not taken. The technology in some respects drove the detail of people's lives, however, and was the dominant force in the strategic vision of the company. In terms of relationships within the company, the things spoken about and difficulties perceived, however, technology was barely mentioned. It was taken for granted. For a company at the raciest end of the high-tech business this was not a problem. Staff expected, indeed wished, it to be that way, as P, the senior technical manager, made clear:

*Interviewer*: I get the impression that you develop the systems, you're pleased with the systems too, by the sounds of it, but nobody sat down and said 'Do we really need this kind of system, should we go back to pencil and paper?'

*P*: I think everyone in the company recognises that would be a retrograde step.

*Interviewer*: So in a sense it was on.

*P*: We're all technocrats, so you can't really expect me to say that kind of thing.

*Interviewer*: Sure, sure.

*P*: Were we a firm of accountants it may of course be an entirely different thing. But we are proponents of technology, and therefore we have a PC on your desk. It depends on the sort of person you are. It's almost a foregone conclusion that unless we use the systems that we sell, unless we exploit the technology, how can we possibly believe in it wholeheartedly?

P was moving beyond the importance of the technology as a valuable tool to a recognition that technology formed part of the way Unipalm staff defined themselves and the company. The culture he was invoking, involving definitions of self – 'We're all technocrats' – and the company – 'we are proponents of technology' – went beyond operational routines and identified the meanings which lay behind them at operational and strategic levels.

*Strategic practices*

The different cultures at Unipalm and Pipex manifested themselves in the taken-for-granted practices of the company. To give a sense of how this worked in Unipalm-Pipex, and to gain an insight into its routine strategic practices, requires a story-telling approach which does more than give the thought-through interpretations of the researcher. The company lived change. Its style and responsiveness are part of the research evidence. In this section I begin by describing events as they appeared to me during one set of visits to the company. Those events encapsulate the culture and routines of Unipalm-Pipex. They are a small number of events, taken from a wide choice, and give the essence of the Unipalm-Pipex world. Van Maanen (1988) would call this an impressionist tale.

The meetings took place in July 1995. I visited the company on consecutive days. On the first day I had meetings with D, the company founder, and K, the junior sales executive; on the second day with H, the assistant to D, W from corporate sales and P, the senior technical manager. The company was expanding fast. It had doubled its number of employees since my previous visit at the end of January. A new building was being opened in a nearby location and the merging of the distribution company with Pipex was under way.

The company's offices were on the Science Park in Cambridge. This is a prestigious location and the architecture is modern with high-tech design predominant. In general the buildings are low-rise with car parks and small gardens dividing them. There is an atmosphere of modernity, advanced technology and youth. It was a very hot day. The grass outside was scorched and everybody was working in shirtsleeves in the air-conditioned building. D came down to meet me in reception and took me upstairs. The building was busy. People's desks were pushed up against each other. D's room was a glass-partitioned office in one corner of a larger room covering perhaps half the first-floor area of the building. The room was relatively small, big enough for a desk and table and a few chairs. Nobody else in that area had a room. It was completely open plan. D could see, and be seen by, the people working all around him.

D was welcoming and very open. He seemed even more full of energy than usual. He spoke using phrases and short sentences, as if paragraphs were unnecessary or old-fashioned. This style of speaking was adopted by many of the people I spoke to at Unipalm, particularly men, and could well be an industry affectation. (It was not adopted at DEC, which was in many respects in a different business with its focus on hardware and associated consultancy.) It gave a slightly macho, urgent and knowing edge to any conversation.

I began by asking what had happened since I last visited the company. D explained that his partner had gone and that he, D, was now Managing Director of the whole group. This had been something of a watershed. The group had switched its focus away from products to particular market sectors.

The conversation moved to the difficulties of managing these changes. D's manner remained relaxed but with lots of nervous energy. He seemed to want to talk about the issues concerning him. He had split up the Pipex sales team the previous week and integrated it on a sectoral basis with the Unipalm team. He used the meeting with me as an opportunity to get things off his chest. 'Well basically I pretty much forced the issues with them. And where there were missed decisions I faced them. Had them in my room, talked to them about it, and say "Well. What have you yet to do?" Rather than say "OK, this is what you must do." On a couple of occasions I have had to just tell them. You know, "You've got to do this." And there was an occasion, couple of days ago, about a week ago, a sales manager rang. The integration, the splitting of the Pipex team was the issue and basically I lost patience with the softly, softly approach. It had been six months and they hadn't integrated properly and so I lost patience and said "I'm a Director. Do it." The Director of Sales was really upset about it on the day. Cut the legs off him. "Completely eroded my authority" and then sort of after a couple of days he confessed, "I should have done that, shouldn't I?" It was time to grasp the nettle, he'd had an unwillingness to grasp the nettle so he is learning that these tough decisions have to be grasped, and take risks, and understand that it's down to him. But, you know, occasionally you have to do directive management, I had to do directive management.'

D had claimed previously that he wanted to delegate more. Going over the head of the Director of Sales irritated him. He did not want to do that, he claimed, but was forced into it. But this was not good practice.

D:    Yes. I should have talked through that before I put it down to his managers, I'm sure. But it was one of those things that, sort of, one of my little. . . . I occasionally get into a mood, which I call Action Man, and it's awful in that I tear up all the rule books and just kind of do things. You know. Unfortunately when I'm in that mode I don't actually think about it. I've thought about what the actions that need doing. OK, I've got frustrated that they haven't happened, right. When I go into Action Man mode I don't, I say 'Well that's got to happen, damn the consequences, what's the quickest way of doing it?' OK? I don't think about, OK I'm going to do this but at least tell the manager first, sort of stuff. And it trips me up occasionally.

*Interviewer*:    But you're learning too?

D:    Well, I wish I was learning but. I know I shouldn't do it but when I'm in Action Man mode I still do it. Grrrrrrrrrrrr.

This was an extraordinary meeting. D had met me several times and saw me in some respects as a confidant. He seemed to find the process of being interviewed relaxing and status enhancing. He was literally acting out his 'Action Man mode', growling and expressing his frustration at his colleagues for not integrating and at himself for dealing with them in a way he regretted. The atmosphere felt electric. I wanted to hear more about his sense of how he should behave and how he did behave:

*Interviewer*: As the company grows, of course, it's going to become more of a problem for you, I suspect.

*D*: I get more frustrated at things going on. There are times when I think 'Well, what do you have to do?' And that can be important little things. I have to complain OK. Network show, big stand, good position and as it happened I went there at the opening of the show. And I looked at the stand and the only words on it were Unipalm-Pipex. Now that was the biggest sin on the stand. You should have your product, you should have [inaudible word] on your stand, with the signage. There was nothing at all. I hit the roof. That was particularly galling because when we discussed the stand I know I said 'I don't care what the stand looked like, I don't care what it costs, but I want internet in six-foot high letters.' And I want people to see, when they go walking in and ask 'I want to do internet, there it is.' And they didn't do it. And I also found out that the Head of Sales and Marketing had said the same thing, yet they'd still not done it. And there are occasions when I think 'What the'. I get really frustrated. And I get doubly frustrated because I shouldn't be doing that. That is well below my horizon, or should be. And uses up so much energy. And occasionally I lose confidence in our success. You know 'What can I do?' But I do. When I see something like that I get such a big stick.

*Interviewer*: You said last time you were, found it quite difficult to delegate and that you felt that you maybe had a reputation for not delegating, even though you try very hard to delegate and believed that you delegated.

*D*: I try desperately to delegate. But my impatience and what have you gets in the way of it. I think. I keep on telling people, 'Don't come and ask me what to do. Tell me what you want to do. Get affirmation', but they don't seem to believe me all the time. It continually gets better. I go on about the company but it continually gets better.

D had now relaxed. The tension had drained out of him and he started to stand back a little from the specific things which frustrated him: 'Fundamentally we have to reinvent the company every six months, changes in market place and so on, changing size, what have you. All but a bit of it is changing. Well a bit of it changes all the time. And quite often people think that. We're a relatively open company and we're open in our thought processes, and so we've got this awful rumour mill going. If you ask a question the rumour mill makes it policy. So, you know, you say what if we turn the air conditioning up? All of a sudden, you know, sort of, the place is going to get a freezer. Or you get complaints for saying that we're already cold enough. Don't, you know, that sort of silliness and that's really quite rife. If we could get some of the positive messages of the company around the company as fast as some of the rumours, you know, it would be amazing. And, of course, if I ask a question it's definitely policy. You know it doesn't confine to me. . . . But. If I could get rid of some of the downsides of Action Man mode but OK you want to do something, you talk to a manager first. That's all I've got to do, and Action Man – to get through some red tape, if you like – that's what you have to do, you have to occasionally, you just have to change gear and push through, OK. And it's, it is not the wrong thing to do, it's just that my character makes it that, it's more painful than it should be. I shouldn't be Action Man all the time, 'cos then I would be doing everyone's job. And you've got to allow people, got to allow people to make mistakes. It's this thing, you don't allow, can't allow people to make mistakes on an ongoing basis.'

This conversation gives a sense of change, the way it was managed and the way in which it manifested itself at Unipalm-Pipex in a manner which a simple description could not. Change was occurring continuously in the company. A routine of sorts was evolving – a hard-driven style with little self-reflection. The rate of change was uncomfortable and the ability to live with discomfort was an important part of routine behaviour for all members of staff during the period of the research. An aggressive and rapid response focusing on the main strategic agent is in many respects an absence of a developed routine. It is simply a working through – almost a caricature – of an entrepreneurial style.

But the Pipex side of the company had begun to adopt a pattern of behaviour which also had energetic and aggressive characteristics. The integration of the aggressive Pipex team with the more structured distribution team created problems. The meanings understood by Unipalm staff did not match those of Pipex even though Pipex had only a relatively short history, and staff had been employed for a relatively short time. A mythology surrounded Pipex, partly because of its newness, and partly because it had been instrumental in the creation of a completely new industry. However, at Unipalm there was a belief that Pipex was in many senses simply another product – not an evangelical operation nor an inexplicably complex piece of technology. W, a senior corporate sales executive with Unipalm, described how the Sales Director of Pipex made decisions about the sales portfolio without consultation and how Pipex sales staff worked in competition with their colleagues from Unipalm: 'but at the same time to dump the traditional money spinners before you have to, I think is really dumb. The Sales Director of Pipex is really looking hard at saying "Maybe we should just scrap all that software." You know the stuff on there's fourteen million last year. You know "Let's dump this three million." He's saying "Well maybe we should pre-empt the market dive and just dump the stuff." This is making five hundred thousand pounds a month gross margin and they're talking about dumping it. People are still buying the stuff. You know, "Microsoft will wipe us out in August." What we've got to realise is that the cost of the product is actually a tiny fraction of the cost of ownership, and Microsoft are giving it away free next month with Windows 95, but that won't stop people buying it. You ain't going to kill this sort of size market overnight 'cos one player intervenes. . . . And so to admit defeat before you've got involved in a battle is stupid. . . . I mean we dropped three products last week. Sales team weren't consulted, just drop them. Now they were like integral parts of some solutions we were building at the moment. Fifty grand's worth of business. . . . That sort of decision's I think is critical. I believe in the future we'll be into solutions but we can't dump all that stuff yet.'

The merger brought a number of the implicit features of the different cultures further into the open. The Pipex approach was beginning to infiltrate the whole company. W had decided to leave the company and described how friction over the merger partly prompted his move: 'There's been like a massive amount of friction. I mean that's one of the reasons I'm going is that I got fed up with the friction. . . . they're supposed to work with us to sell Pipex. Fair enough. So what they've done instead is they've not worked with us, they're trying to put something in. They've been selling products. . . . They've become head to head with us price wise, you know, quote against us, you know the sort of things they should never have been involved in in a million years. They were doing it and we were aware of, you know, situations where the customer said, "Oh such and such was talking about that product as well." "Is he really? Did you ask him if he was account manager?" "He said he was the account manager, definitely said he was the account manager for me." And then you know you sort of ask these people, "You been doing any dealings with such and such?" "No." Interesting. You know. But at the same time you've got to try and work with them, so what do you do? Go and tell your boss "Oh by the way, such and such is lying to us." I mean you just run with it and accept the fact that if he does sell anything into that account you'll get paid for it anyway. So it's all seems to be the same arc, same little sector so you know what do you do about it? Do you get his boss to come down on him like a ton of bricks and therefore lose him as agent or do you say don't do it again? Well it's a debatable situation and I don't know who orchestrated this method. . . . '

These were heartfelt comments from an experienced member of staff who left the company partly as a result of the friction arising from the merger of the two businesses. The aggression of Pipex and its approach that almost anything goes sat uneasily with the more ordered approach of Unipalm. The decisions to 'dump all that stuff' were taken by managers originally in the

Pipex side of the company who had little sympathy for the traditional selling patterns of the distribution business. For them the future lay elsewhere: an internet angle had to be included in every deal. And they brought the aggressive approach, which had been successful in establishing Pipex in the external world, into their relationships within the company.

Thus a routine had begun at Pipex which felt very uncomfortable to those accustomed to the more conventional approaches of Unipalm. It was recognised as the way Pipex did things. They coped with change by moving fast and hard, pushing objections out of the way. It was not a legitimated routine arrived at through debate or notions of appropriate social practices, as collaboration and networking had been developed at the National Extension College, partly because it was seen as a necessary evil rather than a socially desirable set of behaviours, but it was clearly the way Pipex staff tended to behave.

## The evolution of a strategic routine

What we have here is a strategic routine which has a number of components. Firstly, it is important to remember that the extraordinary behaviour described by W and D was built on secure operational routines developed in the Unipalm side of the company. Secondly, the strategic practices which were developing appear to have two parts which reached back into the day-to-day practices of the company's employees. The first part is the behaviour which 'cuts the legs off' the Sales Director or, when carried out by the Sales Director of Pipex, destroys the solutions being developed by Unipalm sales executives. This behaviour is not simply that of one person alone. In the examples quoted here, it is that of D, the Sales Director of Pipex, and the Pipex sales staff.

The second and more important part is the activity underlying this behaviour. Staff are leaping into new areas, almost as an act of faith, and dropping old products. There is a constant concentration on new ideas and new products to the detriment of older areas. This underlying activity need not necessarily, in principle, be associated with the particular aggressive behavioural traits, though at Unipalm-Pipex it did appear to be. The leaps are based on rapid judgements about future potential. In the short history of Pipex such judgements had paid off, thus reinforcing the legitimacy of this rapid forceful approach.

The third component of the strategic routine is the privileging of one part of the company over another. This had been apparent at least from the flotation of the company. Action Man and product selection were derived partly from this. They drew from the culture and relationships in the company established by D and in some respects reflected his personality.

I sensed that part of D's agenda in his conversations was to establish himself in his own mind, and mine, as a successful manager and entrepreneur. Revealingly, in an interview after he had left the company, he stated that

Unipalm-Pipex could now never be seen as failing because it had disappeared as a separate unit, and that he had come out of it with his reputation largely intact. These were his reflections on success – a fear of failure rather than pleasure in success. An article in the *Guardian Online* also reported him in this light under the headline, 'UK net loses pioneer':

> Pipex, the main provider of Internet access to UK businesses and a growing force in the dial-up world, last week found itself without its founder, the man who shaped the UK Internet scene for the past five years.
>
> [D], 43, resigned as European manager for Pipex . . . [D] says his reason for leaving is quite simple. 'The business of being an ISP terrified me', he says. 'And eventually the sheer terror palls.' He had grown 'uncomfortable about managing such a large business' and decided to get out.
>
> (*Guardian*, 23 May 1996)

This matches an attitude which has been observed elsewhere. Andy Grove, the co-founder of Intel, and Bill Gates, who founded Microsoft, have little in common except that, allegedly, they are professionally paranoid. Grove maintains that only the paranoid survive (Schofield 1996). Fear of failure is consistent with the emerging Unipalm-Pipex routine.

### The modem story

So far the focus has been on internal relationships in the company. The company's strategic routine also drove change externally. The company continually tried to keep ahead of its rivals and in some cases changed the perception of the external world about the nature of the internet business. In the same set of interviews in which D described his Action Man mode, P showed the tension he felt in creating a completely new business.

P was a senior technical manager. He had wide experience in the computing industry. On the day following the conversation with D, I met P in an animated state. He was wearing casual clothes instead of his usual suit. P always struck me as someone who was concerned about his appearance and his clothes were usually fashionable, so this was a surprise. The weather remained exceptionally hot and he looked uncomfortable. The company was expanding into a second building. He had been instructed to arrange major changes to the cabling that had been put into the building. P was quite literally getting his hands dirty. The new building was to be used for Dial, the Unipalm-Pipex product which brought the internet into the domestic and single-user market and which was a brand new area of work for the company. P had been given responsibility for Dial a few weeks earlier. P's manner was urgent and abbreviated, rather like D's. In my notes of the meeting, I described P as a personification of the company culture. I asked him why the company had moved so fast into the single-user market when six months earlier they had claimed no interest in it. I have set out the transcript of that part of the conversation below. Again, it gives a sense of what the company felt like and adds to the conversation with D. P describes the process of moving into the single-user market with the Dial product (initially called Solo) in an intense, riveting way:

*P*: we've got the software product. We've got the demand, got the market place and it was very clear to us with our visibility in the market the demand was going to come. We've talked so far about the differences in dealing with the corporate market place and the mass market, so on and so forth. We've always been geared up to handle the corporate market place. We're not geared up to handling the mass market, so we had to gear ourselves up pretty quickly, 'cos the volume of enquiries that we've seen were going to create a lot of noise and a lot of distraction and a lot of irritation for us. And a term that I've used on a number of occasions is you know if we were going to be dragged kicking and screaming into that market place whether we liked it or not, then we could either do it under our own control or we could do it out of control. And so we had to turn our attention to it and react fast. . . . We have eighty per cent of that market place today in UK corporate connections market and if you extrapolate the growth, the projected growth in the market place and what we're planning to do we will actually lose market share by growing as fast as we possibly can. So there's more business out there, enough there for everyone. But we don't want to lose out too much. We've really got to strive as hard as we can to keep pace with that. Couple that with the fact that, yes, we can sell what it is we have today, usually what we have today isn't what we're going to be selling tomorrow. An opportunity comes along. Experience has told us that if it's not something we've done before we still need to look at it very very carefully, because another two or three will be along in no time at all. . . .

*Interviewer*: Do you feel that if you didn't grow so fast and keep up that you might be swamped and just disappear? Is that the sort of threat? It's either keep up or disappear? I guess it is exciting.

*P*: It's frustrating, it's exciting, it's annoying, it's irritating, it provides you with adrenaline, it becomes the reason why you get out of bed, it becomes the reason why you couldn't sleep when you were in your bed, it becomes all-encompassing.

*Interviewer*: I was going to ask how specific changes in technology might have affected this and I guess you in a way you've answered that. You're basically trying to keep up with whatever's going on in the technologies.

P:          Yes and one of the problems is in trying to move forward many
            areas of technology whilst still addressing the growth and the
            new opportunities that clients are throwing at us with the
            solution to creative takes an awful lot of skill and an awful lot
            of time. Having anyone left who can pick the phone up and
            say 'Yes Mr Customer, I'll sell you one of those today' is hard
            work. 'Cos everyone's ricocheting off the walls.

Interviewer:   So in a way, I'm not trying to put words into your mouth but
            just thinking aloud really, in a way the structure of the company
            is almost defined by the fact that it's in this particular tech-
            nological area?

P:          Yes, very much so. I mean the demands of the clients are
            actually creating this Dial, corporate and international split.
            And it's quite conceivable that the Dial piece may actually split
            into two. I can see that, personally, in the future as well, not
            too far away. Maybe before we've got the whole team built.
            You have to be prepared to flex the structure of the company,
            flex your plan for the structure of the company, in response to
            the demands that you see.

This is growth and change generated by the external world. There were new
opportunities, which were hard work, and future changes in the offing. 'You
[had] to be prepared to flex the structure of the company' to respond fast to
change. But it seemed to me that perhaps P wanted to be seen as an important,
go-getting manager. Perhaps some of this drama was his fantasy. I wanted
some examples of why 'everyone's ricocheting off the walls'. His answer set
out a dramatic change.

Interviewer:   What typically would be the sort of notice that you have
            of something like that? I mean let's think about Dial. I mean
            when as you say when we were here last speaking, Unipalm
            wasn't in that market at all. That was different. Now it's a third,
            at least a third of your structure if not a third of your business.
            Talk me through how fast that happened, what, how come,
            you know, six months you've suddenly taken on this main new
            area. What?

P:          We produced a product at the tail end of last year called Pipex
            Solo, which was, as its name would imply, a single-user dial-
            up connection to the internet. The intention was, bearing in
            mind that we were almost entirely in Pipex area, and I wasn't
            involved at all in Pipex at that time, not at all. I mean Pipex
            was those guys downstairs in inverted commas. Maybe
            not quite that step but you understand what I'm saying, my
            day-to-day activities, it was outside my field of vision. But it
            was intended to be used in the corporate market place for two

reasons. One is as a taster for the key decision makers who we were trying to influence, who couldn't even spell internet but they'd heard of it but didn't really 'My IT guy tells me that we want this. Convince me.' And the other was for those people that had taken the corporate connections to use for teleworking. Now that addresses two nice areas in the market place. One as a hook for those people to take on the eleven thousand pound a year connection and the other into teleworking, very much an area of interest. So we produced this product called Pipex Solo. Suddenly it looks very appealing to everyone from the key decision maker in British Gas at board level right down to little Jimmy in his bedroom with his Amstrad PC and a modem and his mum and dad haven't got a clue what he's doing, but he's certainly on the net, sending e-mail to his girlfriend, OK? And that product, by accident rather than design, fitted those roles perfectly. Nothing else like it on the market place, time was right, the market was right. Part of the marketing machinery meant that *PC Answers* magazine caught onto this and put a cover disk on their magazine and it hosed our network. We were in the middle of some upgrades, so we weren't too worried about the extent of that and fairly shortly after we did one for the *Internet* magazine, a cover disk, and again it just destroyed our network through sheer volume of demand. So then that was the point at which we said 'Look we're getting lots of these enquiries, we're getting dragged into this, we've got a product, it's here now, we need network infrastructure, we either disappear into complete and utter chaos saying "We can't cope with this" or we put our minds against it and we say "We can cope with this by doing this, this and this."'

Rather like D the previous day, P was animated. He was describing actions taken by the company more or less to test the water which 'hosed our network'. The network staff and the marketing and sales people were screaming. How would the company react? What would the routine response be? The company had taken a small step and suddenly found itself in a new world. What could it do? P continued: 'Part of that was [D's] command. His command from on high to me "[P] do this." And I built the network. On the day that he gave me that I had no idea what to do. Didn't know what the technology involved was, what the modem racks looked like, what bits and pieces I'd need, what Lego bricks I'd need to put in place. Where the suppliers were, how it all fitted together, what space we had. We didn't have space to put it anywhere. Just knock a few walls down, move a few things around and it all comes together. But, you know, [D] and I put our minds round it and we came up with an attitude of mind that said "We can do it."

The only things that'll stop us are things that'll get in our way so we move them out the way. . . . Well I mean it was a problem. It was a problem, it was impacting the network and we took a decision that says "Right. We go for this market place." The marketing people were saying, "Look. We've got this cover disk but there's another half dozen that want to do it. The opportunity's there." There's a lot of sales people saying to us "Well, resellers want to resell this product. It's really nice, it's caught the interest." You've got the makings of a business there. You've got a market that demands a product. You've got a product that's there and deliverable. All you need is a network and a structure of people to be able to address it. All the key elements, the basic building blocks for a successful business were there. All you've got to do is resource it.'

There was then a brief discussion about the main competitor for Dial, a company called Demon, and P continued with his explanation of the Unipalm-Pipex *volte-face*:

P:             . . . And there seems to be a steady migration of Demon users to us. If that stops then maybe that's one of the danger signs. I don't know.

*Interviewer*: Why, why do you think that's happening? I don't know anything about either product so. . . .

P:             It's the capacity of their network. And we've invested a lot of money. I spent two million pounds, since you and I last spoke, on deploying this. Two million pounds that we didn't intend to spend back in January. Brave decisions! And we have enough modems to ensure, and enough intelligent call routing as well, to ensure that users don't hit an engaged tone. And that is the main bugbear for Demon users.

P described these changes with a sense of pride. He had been given a task far bigger than he had expected in an area new to him and it had gone live. The big investment had been in network capacity. Network capacity determines the speed of response which users receive when they dial in to their internet service provider. It is a major measure of quality. Unipalm-Pipex seemed to have got it right. 'Things that'll get in our way . . . we move them out the way'. They were able to upstage their main competitor because they could give a quality response. What I had not realised at the time of the conversation with P, but was revealed to me much later in a conversation with D after he had left the company, was the sheer scale of the investment Unipalm-Pipex had made. I'll let D tell it in his own words:

'there was no-one spare apart from [P] so I went to [P] and said "Go and get more modems, modems in four–six weeks. How quickly can we do it? Find out as fast as you can." So anyway we came out of that, five hundred new modems in about six weeks and he came back after two or three weeks and said right "Well I've got two hundred and fifty modems ordered and

I've got two hundred and fifty lines" and I was so cross. "I said five hundred. And when you've got those five hundred go and get another thousand." And all of a sudden he thought "Ah. You're not kidding, are you, about this?" It's a case of if you're going to do it, go for it. OK and he went off and got five hundred, got the thousand, didn't need the whole thousand. Cost us a million pounds that mistake. It actually sort of points to the expenses. It cost us a million pounds and probably made that business. There's a lovely post-justification for it. Because he'd gone, the key thing is that those events, [P] all of a sudden understood that we were in this business, you know, we weren't playing around. And we've got to make a success. You know, you make the decision and you've got to make it work. Of course one of the consequences of that, going from a hundred and twenty lines to sixteen hundred is that you actually need ten times more customers. So all of a sudden it was no longer a problem of lines, it was a problem of customers, we had to go and get customers. 'Cos the money was going out the door. So that that those events kicked [P] in. He got belief in himself, he actually did it as well, OK so he got belief in himself, a nice self-fulfilling, up the spiral for [P]. And the other interesting thing though is that because [P] went in for five hundred and then asked for a thousand all of a sudden the Telcos took us seriously in this business. Mercury, Energis, BT what have you all of a sudden they started talking to us about discounts and you know. Occasionally people come for a five hundred order. No one ever goes five hundred then two weeks later asks for another thousand. Doesn't happen. So all of a sudden it kicked over the Telcos into realising that it was a serious business and that Unipalm was a serious player. And that goes for the modem manufacturers. All of a sudden we went from being a customer to being a strategic partner.'

The last point here is very important. By expanding ludicrously fast by industry standards – buying 1,500 high-speed modems at a cost of several million pounds in a very short time-scale, Unipalm-Pipex changed the perception of the industry. It enacted its own environment. The Telcos (industry-speak for large telecommunication companies) such as BT, Mercury and Energis, suddenly took notice and the world had changed. This is symptomatic of the Unipalm-Pipex approach. D's sense of the market plus his rush of blood in Action Man mode was just right. It could have destroyed the company, but it transformed Unipalm-Pipex from seeking capacity to seeking customers and made it a major player in this market. The strategic routine involved creating a decision from which the company could not back down. Changes in the market were analysed fast. The analysis was then put into operation at a big scale. P called these 'brave decisions'. If the analysis was correct, the outcome, as in the modem story, was a success for the company. If this pattern of jumping in with both feet had been wrong, it is likely, in this case, that the company would have failed. This is a formative practice of the strategic routine: make decisions and take irrevocable steps. As D said, 'It's a case of if you're going to do it, go for it.'

P did a good job. At the end of the research period he was the only remaining contact still working for the company and was in charge of the Dial project. It may well be that his success with Dial, and the initial frisson created for the Telcos when he bought the modem capacity, were influential in making MFS (one of the Telcos) take an interest in the company and propose a merger. This was never suggested but it is plausible that the change begun by the modem purchase resulted in the large telecommunications companies taking the new internet service providers much more seriously.

### A routine?

Well-defined operational routines existed in the company. Action Man mode and the modem story tell us about emerging routines at policy levels in response to change and as a generator of change. There is a consistency of behaviour and a regularity which reveals itself in many aspects of the company's practices.

Unipalm-Pipex was successful. It had moved fast, generated profits from distributing software and a rapid increase in share price on the back of internet service provision. It was becoming a major player in a market which, apparently, lots of people wished to enter. It appeared to be a self-confident company. What it had done had worked and it continued to operate in that way. If the company had been unable to incorporate change, it probably could not have survived. And the way of doing things was an aggressive, go-getting style, driven by a fear of failure.

The modem story tells us how decisions were made fast. A cover disk on a magazine took off unexpectedly well. The implications were rapidly assimilated (by individuals who had a record of apparently successful judgements and a particular world view). The response was to force change by developing the product extraordinarily fast, involving steps which committed the company to the change. It could not back down.

This routine had evolved in the company and by the time of the modem story was certainly established, if not entirely sanctioned. It enabled the company to incorporate change. The underlying rule appeared to be: react very fast and at scale. The practices which followed from this rule were the aggressive, sometimes abrasive behaviours, with little self-conscious reflection, outlined in the stories.

Strategy was developed in other ways too, at Unipalm. The stock exchange flotation involved many meetings, debates and presentations and was more thoughtful, but in responding to perceived market changes and opportunities the strategic routine outlined here, and exemplified by two stories drawn from the fieldwork, typified the company's strategic practices.

We cannot know how long the company could have continued if it had not merged with UUnet. The sheer energy and stress involved in operating in this way casts doubt on the routine as a long-term pattern of behaviour, but that does not mean that change cannot be incorporated long term, as the next chapter will go on to show.

The routines (developing and actual) were rooted in the history and culture of the company. The company's structure (and changes in its structure) and the changing market, plus the skills and competences of the main agents and their search for security, were also implicated.

## Conclusions

Operational routines embedded in the technology and linked to standard patterns of distribution were ultimately adopted throughout the company. The operational activities of the company were securely tied into the electronic systems the company used, and there was a reassuring predictability about them. They were changing but in ways which were built on previous experiences. Operational memory, and routines embedded in the software, changed slowly as the software and needs of the staff developed. This was a stabilising force within the company and allowed it to go about its day-to-day business effectively. It was based partly on an industry recipe concerned with sales of products which are technically complex and require customer support.

Developing a strategy for the internet side had no obvious recipe. The internet itself has been viewed as a network which opens up new horizons in which people are not heavily constrained. Attempts to control the internet by commercial organisations or governments have been strongly resisted by the original 'cyberfreaks'. Internet people have an image as anarchic, enthusiastic, technologically obsessive and sometimes arrogant. This image idealises the individual and particularly the 'wacky' non-conformist who is keen to do his or her own thing. Pipex had something of that culture about it and in a broad sense matched the wider image of the internet. Changes frequently resulted in pain for individuals, and discomfort became institutionalised. This was part of the incorporation of change in a turbulent environment.

The technology was a constitutive element of Unipalm-Pipex. P has already set out this interpretation. He made the point more indirectly but perhaps more tellingly in a different conversation, though on this occasion he was talking about the whole Unipalm-Pipex group after the merger with UUnet, not just the Unipalm side: 'So how you convey the mood and the meaning. . . . Electronic mail has scaled very nicely with the company. We have electronic mail that is as good and effective as when the company first started using it, with about six employees. People use it naturally. There are some people in the company that don't actually talk to anyone any more, they just stare at a 17"dome or piece of glass all day and that's fine. You can actually be very productive. It allows you to commun[icate], use one communications medium to have an interactive conversation with somebody, get your answer back two minutes later and you bounce another comment back two minutes after that. Threaded, inter-threaded through the course of the day, you can have a complete conversation. Or you can leave a message for someone. They don't have to be there to take that. . . . All of the members of the intelligent team's here, so it's a good way of communicating at a team level. Now I don't drive the team by the day. . . . I sit at my desk, and I can be having a conversation with you and mail is coming in all the time and those that are aimed at consultants I can see reasonably easily. So whenever there's a quiet five minutes I can just go through this, and I'm up to speed on everything that's going on.'

P was describing these features in what was then UUnet-Pipex. He demonstrated what he was saying from his own desk. There was much more of a big company feel. P had his own room. His 'Dial' team were in the adjoining area. He pointed proudly to one customer adviser who was practising juggling while talking to a customer on his hands-free phone. The culture was shifting but at the heart of it was the technology that made it all possible.

The company thus focused on technology in a deep sense. Technology formed part of its unquestioned way of doing things. Informating, implicit in P's final comments above, was taken for granted and operational routines were embedded in the technology.

Orlikowski's distinction between technology as artefact and technology in use is revealing (Orlikowski 1995). The technologies which P was then using were commonplace. E-mail systems and advanced telephony were no longer the state of the art. The product enhancements he used to attract custom, he admitted, were not sophisticated. Value was added for customers by giving multiple e-mail addresses or more web space in the standard package. But the technology was everywhere. It was pervasive and part of the UUnet-Pipex world. The social and the technical are thus inseparable. They are different units in a network (Latour 1996b; Walsham 1997), which through their relationship with each other and the organisational and social structures in which they are nested, interact in the reproduction and transformation of routine behaviour.

The strategic routine adopted at Unipalm-Pipex was a fast-moving, aggressive response based upon an immediate assessment of the relevant circumstances and a belief that it was important to move faster than the competition. The company continually took decisions which redefined the environment and moved it into unknown, and unknowable, territory. The routine was adopted by managers throughout the company. It was the way things were done and taken for granted, a set of strategic practices in the company. The routine is not the same as company culture. Culture establishes shared meanings (Hall 1997), and is deeply implicated in practices, but the two are separate.

The strategic routine arose from a range of factors. The culture drew partly from the personal style of the key agent, D. The rate of change of the environment and the insecurity this created in the company's staff were also influential. The routine seemed to work and it provided a route map through uncertainty. The modem story is a rich example of this and also illustrates well how the routine itself changed the environment: Unipalm-Pipex suddenly became relevant to the major companies in the industry. In the modem story the company responded fast to the unexpected demand for its new product. The opportunity was perceived and followed through. D saw the need to enter on a large scale, on the basis partly of his knowledge of the market place and partly of his Action Man style, which had become part of the company's way of doing things. His style had become locked into the routine. It became difficult for him to operate in any other way. People would not come to him for 'affirmation' (the word he used in discussing difficulties of delegation) of their proposals so that he had continually to determine what went on. Other staff also operated in similar ways to D. P drove the modem changes through but expected the environment to change and was continually looking out for new and unexpected events. This produced the adrenaline which got him out of bed in the morning.

Internal company structures and relationships also played a part in developing the strategic routine through interactions with each other and the external world. D's overriding of the Sales Director in breaking up the Pipex team and his discussion of the rumour mill exemplify this mechanism: his relationship with the Sales Director illustrates a company structure which was loose. Little respect was paid to established reporting lines. As a result rumours became extensive as a means of communication, partly because conventional reporting lines seemed unreliable or contradicted the apparent direction of change. These things were happening throughout the company, however, and were not peculiar to D. The Sales Director himself reportedly took decisions which rode roughshod over his staff. The company was not anarchic nor was it driven by D as an archetypal, single-minded entrepreneur. Both those images – anarchy and single-minded dominance by an individual – miss the sense in which this company hung together through the co-operation, and competition, between its members, running all the time to survive by staying ahead of the rest of the market.

The strategic routine then incorporated change. It did not respond to change nor reflect on change but took change as the norm and in doing so was itself implicated in many of the changes facing the company. Structures and norms, both external and internal, changed as the company or staff members interacted with them. There was a groping towards a regular pattern of actions as individuals within the company began to impose meaning on the swarm of events.

As the company grew, and ultimately merged with UUnet, the interpretations became more sophisticated. P's comments in my final meeting with him, several months after the UUnet merger, indicate how some of the immature strategic practices of Pipex had matured: 'There was the flotation, there was – this is all in the last two years – flotation, we merged the Unipalm and Pipex halves of the company. We built the Dial business and restructured into two key, well, three key, core trading parts of the organisation. We've merged with UUnet, American company, which again was more change and now, sitting on that bulletin, is the next change. MFS and UUnet have agreed merger, so in September it's "As you were lads, let's throw all these coins back up in the air again and see which way they land now." And you actually become quite adept and accustomed to working in that kind of environment where you can say "Well who do you report to?" "Does it really matter?" Does it really matter, because it's going to be someone different, you know, and you're constantly trying to build relationships not to the point of necessarily an employee with his boss in the traditional sense but as a colleague who, you could easily find the person who is your boss today working for you in six months time or working alongside you or in an entirely different part of the company.'

P is showing how the expectation of change had become institutionalised.

At Unipalm-Pipex the operational routines were secure. Beyond the narrowly operational, we can observe a regular pattern in the approach of the company to the changing world it faced. Looking for such patterns helps us to understand the firm's behaviour. My initial image of Unipalm-Pipex was opaque. It was difficult to know what was happening and why. The staff interviewed also professed difficulties in keeping up with what was going on. However, searching for patterns in working practices, which I have argued are sufficiently regular and sanctioned to be seen as routines, enables us to see how apparently diverse behaviour fits together. The world of Unipalm-Pipex, like the other case-study companies, was constituted by a recursive understanding linking structure, agency and technology and external frames of meaning or recipes. Recurrent strategic practices arose from this which, in this case, were able to incorporate change. The existence of such routines is not peculiar to Unipalm-Pipex, however. The next chapter goes on to show how a company with an apparently very different culture also developed routines which incorporated change.

# 7 Chadwyck-Healey: change, routines and cultural contrasts

## Introduction

A visit to Chadwyck-Healey leaves a completely different impression from a visit to Unipalm-Pipex. The company is in a different industry and has a different atmosphere and style. In a great number of respects, however, the two companies were very similar. Both were operating in a rapidly changing environment in which they had a formative influence in the invention of a new market. Change was driven by both companies, not simply responded to. Both companies experienced a major change in orientation. Chadwyck-Healey was also able to incorporate change into its routine practices but in a less frantic and more thoughtful way. Why was the path followed by Chadwyck-Healey different from that followed by Unipalm-Pipex, even though many of the circumstances it faced were similar?

The contrast between the two companies demonstrates that strategic routines which incorporate change can be found in different environments. Such practices are not unique to Unipalm-Pipex.

The strategic routine to be examined here can be encapsulated in a phrase which was used, in slightly different forms, by many people at the company: 'You're only as good as your next project'. It meant that the company espoused an orientation which looked to future developments rather than resting on existing products and practices. Lying behind that espoused approach were different cultural norms from those found at Unipalm-Pipex but also elements of Action Man. It was also clear at Chadwyck-Healey that the links between strategic and operational routines were important, that some strategic practices would have been impossible without particular operational routines.

The chapter will focus on the development of a new product at the company, called Lion. Through consideration of that product and other developments in the company which were linked to it in different ways, we can see how a primary strategic routine worked in the company and analyse the factors on which it was based.

## Background to the Company

In November 1996, in a news item on its web-site celebrating its winning the Queen's Award for Export, Chadwyck-Healey described itself as follows.

> Chadwyck-Healey is a publishing company which specialises in reference and research publications for the academic community, public libraries and business. It publishes in a variety of media, including CD-ROM, microform and paper, and will launch next month its first Internet service. Its main export customers are university and other research libraries in continental Europe, North America, Australasia and Asia.
>
> Since 1991, when it announced The English Poetry Full-Text Database, Chadwyck-Healey has become the world's leading electronic publisher in the humanities and is converting the printed texts of the past into digital form on a scale unmatched by any other publisher. English Poetry contains over 165,000 poems by more than 1,250 poets, the equivalent of 4,500 printed volumes.

It continued, after a description of some of its literary databases,

> Literary databases are not the only publications which Chadwyck-Healey sells strongly overseas. The company markets, for example, the CD-ROM editions of *The Economist* and six British national daily newspapers, The British Library's catalogue and national bibliography and the British Film Institute's database on films and film-makers, Film Index International. International publishing partners include the United Nations, the Office for Official Publications of the European Communities, the National Security Archive in Washington DC and the State Archival Service of Russia.
>
> Chadwyck-Healey Ltd, based in Cambridge, is a privately owned publishing company established by its Chairman, [Y], in 1973. It has sister companies in France, Spain and the USA. In 1987, when it won its first Export Award, export sales were £970,000. 24 people were employed in the Cambridge office. In the year to June 1995 export sales were over £7.5 million and staff numbers had risen to 102. Since 1993, export sales have increased by 135 per cent. Chadwyck-Healey Ltd now employs more than 150 staff in Cambridge.
>
> (Chadwyck-Healey 1996)

The press release gives a good sense of the company. It is low on hyperbole, changing fast, with an emphasis now on *electronic* publishing. It sees itself as a publisher not a provider of electronic services. Table 7.1 confirms the claims made in the company's press release.

The company was established by Y in 1973 as a specialised publisher of research material for university libraries throughout the world. It is a private

*Table 7.1* Chadwyck-Healey Ltd, company financial profile during the period of the fieldwork

| | 06/96 | 06/95 | 06/94 | 06/93 | 06/92 | 5-year average |
|---|---|---|---|---|---|---|
| Turnover | 9,741,290 | 8,944,655 | 7,747,070 | 5,037,960 | 4,093,348 | 7,112,864 |
| Profit before tax | 666,492 | 1,179,664 | 1,001,307 | 232,671 | −217,867 | 572,453 |
| Net tangible assets | 2,185,161 | 1,907,587 | 1,125,541 | 558,841 | 390,067 | 1,233,439 |
| Shareholder funds | 2,164,143 | 1,807,961 | 1,101,244 | 540,897 | 369,659 | 1,209,380 |
| Profit margin (%) | 6.84 | 13.19 | 12.92 | 4.62 | −5.32 | 6.45 |
| % return on shareholder funds | 30.80 | 63.05 | 90.93 | 43.02 | −58.94 | 33.77 |
| % return on capital employed | 30.50 | 61.84 | 88.887 | 41.11 | −54.03 | 33.66 |
| Number of employees | 132 | 102 | 78 | 63 | 53 | 85 |

Source: FAME Financial Analysis Made Easy Database.

Note
All figures are given in £ sterling except where stated; columns refer to the financial year ending in the month indicated.

limited company. At the time the company was founded the economics of reprinting were deteriorating and a move to microfilm/fiche seemed to be sensible. Y's previous employer was not interested and so he decided to move into that market place himself.

Microfilm/fiche seemed to be an ideal medium. There were no stock problems and the technology was relatively cheap. The initial project was to microfilm publishers' archives plus some parliamentary papers and, initially, to establish the market, the user technology, microfiche readers, were given to customers free of charge.

The major new development in the early years was the publication of the catalogue of non-official British government publications (non-HMSO). The project started in 1979 and first publication was in 1981. In 1987 this was combined with the HMSO database. Now Chadwyck-Healey publishes a complete catalogue of all British government publications on CD-ROM. This became a major revenue earner for the company and permitted the development of subsequent projects.

Y became aware of computer-related possibilities in the early 1980s. He first saw CDs in 1984, and realised that the CD was an appropriate medium for the company and that he should watch for the arrival of a CD market. He saw it as a natural progression from film/fiche. Existing product lines needed machines to read them. There was no change in culture required in the company to move to CD. Similarly the company is used to working on a one-to-one basis with end users. Average booksellers were not oriented to selling CD-ROM products. Much as in book publishing, CDs require a high initial outlay (for coding), then subsequent reproduction is relatively cheap.

The first project on CD – the poetry project – was announced in 1990, orders were taken in spring 1991, first deliveries were in 1992 and the project was completed in 1994. Newspapers were put onto CD-ROM in 1992. In late 1995 initial ideas about a virtual library were being floated. This developed into 'Literature on line', known as Lion, which was unveiled in the first week of December in 1996 and formed the core of the company's development activities. (In October 1999 Chadwyck-Healey was sold to a US-based company, Bell & Howell, for more than £30 million. The company was described at that time as 'easily the most innovative of electronic publishers' (Goddard 1999).)

### Staff directly participating in the research at Chadwyck-Healey

Four staff from Chadwyck-Healey participated in the research:

Y:  the company chairman and founder.
N:  with the company for six years when the fieldwork began, he held a relatively senior position as a Managing Editor. He was responsible for the conversion of text into computer-readable form. The process was

known in the company as 'data conversion'. It involves substantial coding carried out largely by staff in Cambridge and the keying of the coded data primarily by independent suppliers based in India and other parts of Asia;

U:  Sales Manager for the UK and Eire, he had been with the company for over five years at the start of the fieldwork. Chadwyck-Healey had operated with a specific sales force only since 1989 and U had therefore been at the company from the early days of a dedicated sales team. U had worked for a major publishing company prior to joining Chadwyck-Healey;

F:  he joined the company three months before the fieldwork began as Chief Accountant. He had worked in the accounts department of a small company in an entirely different sector prior to joining Chadwyck-Healey.

No very junior staff participated in the research interviews at Chadwyck-Healey. Interviewees were nominated by the company, and it reflects the company's view of itself as a professional organisation (which is discussed below) that only staff in more senior categories were selected. I spent some time talking in general terms to, for example, the Personal Assistant to Y in order to check that my perceptions were not out of line with a broader view of the company but these conversations were not recorded.

## Change and cultural meanings at Chadwyck-Healey

The shared goal of the company was the publishing idea. Without project ideas, it was claimed, the company would cease to exist. The phrase which was repeated around the company has already been commented upon, 'The company is only as good as its next project'. The publishing idea was a goal of the company and the importance of projects infiltrated conversations with all staff. F was newly appointed as Chief Accountant when I first met him. At that time he did not use the phrase. A few months later it tripped off his tongue: 'I think this is true about any business but Chadwyck-Healey's case you're only as good as your next project. So that and this is, I think this is me appreciating what the business is all about.' Y, the company Chairman, used the same phrase in a similar way: 'But it is a strength in that after a time once a project is conceived and on its way, as far as I'm concerned it's in the past. And you're only as good as your next project, as opposed to your last one.'

A lasting image of the company is the concentration on new product ideas and this focus cannot be over-emphasised.

Ideas were developed in an atmosphere of financial conservatism. Y's measure of success was tied into careful financial management. The company was coming up to twenty-five years old, which by the standards of many small companies, particularly those concerned with new technology, is a good age.

Y used the company's longevity as a success measure but explained it largely by the conservative way the finances were managed and by the fact that, in a private company, there was no one else to whom an answer must be given: 'I think the reasons why any business succeeds is difficult to answer . . . because first you have to measure success. I would say that in certain people's eyes you could say that Chadwyck-Healey has effectively failed over the years because it started a very long time ago, I started in 1973 and has only grown by, you know, X amount in that period, when business in such an attractive and potentially profitable area if properly managed and thought through should have been ten, twenty, a hundred times larger, I mean, you can take your pick. We should have gone public ten years ago, we should have done this, that and the other, so, you know, what is success. If you are talking about survival and profitability then sure we have survived through at least three major recessions, the '73–'74 recession, the '79–'80 and the '91–'92 and I suppose that in itself is something. And I think it's basically been a combination of being prepared to take what appear to be risks in choice of products, publications, which appear to need a large investment to get them off the ground, combined with what is actually a very conservative way of running the business, particularly from a financial point of view. We have never borrowed very much money, we have always written down our work in progress fairly aggressively so that we do not delude ourselves in believing that we have assets which are really worth much less than they appear on the balance sheet. And we have never, for instance, put a value on good-will. We've never tried to be profitable, fix, by kind of fixing the figures, because we've never really had an audience whom we've had to impress other than possibly a bank. And we've always, on the whole always paid attention to cash flow. And when we haven't paid attention to cash flow it's usually been, it's caught up with us very quickly.'

This attitude was firmly held and reinforced by the experience of observing other companies overreach themselves and go under. Lion implied a potential reduction in financial conservatism and this will be picked up later in the chapter.

While it may have been 'a very conservative way of running a business', Y continually reassessed his role. He had not always done this but it was clear during the fieldwork, from the way he talked about the company, the books on management which he read, and his use of a consultant as a 'directorial therapist', that he was self-conscious about the role which he played. In conversation, he mentioned several times the way his approach to managing had changed.

'I didn't start off years ago as a people person, as a natural leader or manager of people. I had actually a very negative view of the role of other people in the company that I wanted to set up. Basically it was, I think as you've already quoted, it was me and a bunch of helpers. Obviously I wanted to, I valued them, 'cos even from day one I had one or two extraordinarily able and loyal people, but I am also extremely short-tempered so they got it in the neck. If I was having a bad day I made sure they had a bad day as well.

And it took me years, I mean, more than, I would say ten to fifteen years to really come to terms with the fact that if you assembled a group of people around you to, or the act of assembling a group of people around you to do something was more than just hiring helpers to do specific tasks, the company became, as you added more people you gave them more responsibility. In a way it was no longer such a, just a kind of ego trip. It was, the company became something else. And you might technically own it and have your name on it, but in a strange way it was no longer really yours, in the true sense of the word. And then, as I was able to find very competent people at a senior level I found I was able to begin to relax and not feel that everything was, kind of, resting on my shoulders. And I suppose that was toward the end of the eighties and really suddenly began to actually enjoy the dynamics of having a group of people to manage, and the larger we've got the more fascinating I think it is, and the more I enjoy it. Because if you have over a hundred people and most of them are young, many of them in their first job, I find inevitably I think more and more about what can we do for them, rather than what can they do for us. You know, how can we motivate them, how can we train them better than – at the moment we give virtually no training. What goes on in their minds? What are they thinking about when they come into the office every day? I can remember only too clearly how I regarded my first job and a lot of the time it was sometimes I was very enthusiastic and other times I didn't want to come to work. So I imagine the same thoughts go on in a lot of people here.'

Y had shifted in the way in which he viewed the company. He used to be very much hands-on but became more detached. His role as Chairman was up to him to define. He had no specific portfolio but remained the major source of the ideas the company craved. In this role he had the trust and confidence of his colleagues. All respondents commented on their respect for this part of Y's work on several occasions. For example, U, a humorous and likeable sales person but with the cynicism that goes with the terrain, put it like this:

U:          I think it's partly because we're in a good position in the market. I mean I believe that we could do more in the market, and I think that we have a very, very high reputation, a reputable reputation and I think that goes before us.

*Interviewer*: Do you deserve it?

U:          Well I think that it's, in many respects it's well deserved, yes. I mean for being very innovative, certainly, and [Y] has been very, shown a great deal of foresight in the market and that goes back to the very beginning of Chadwyck-Healey, when [Y] first started publishing.

Y had been dominant in the company for the whole of its existence. It is only in the recent past, by Chadwyck-Healey standards, that the company began to have a sense of identity independently of its founder. Y had a very strong sense that the company was a publisher and this had been important in the way it viewed itself and the industry of which it believed itself to be a part. The following conversation is informative. Y spoke with more determination than in any other discussions I had with him:

*Interviewer*: It's a different feel from that. But it isn't a publisher in the way that, say, Cambridge University Press is a publisher.

Y: I absolutely disagree. I've been thinking of trademarking the term 'The World's University Press' because, except that it would infuriate CUP and OUP who are on the point of licensing things to us. . . .

*Interviewer*: I suppose actually it's interesting that Lion is, as it were, more like a publisher, and one of the reasons that I'm pushing on it is if you look at the, as it were, if I go to an editor at CUP that person will go through something that I write and will, as it were, correct it, and make suggestions about it, but will be changing things, whereas what your people will be editing is actually take exactly what is written and don't change it at all, it's a very different kind of role. They're adding things to it.

Y: Well again I disagree. I mean you're talking about an ideal world. The typical academic text is usually copy edited by somebody aged about eighteen who hasn't got the first clue about anything, well there are some very bright people there but I mean many many texts go through publishers, if you read them they're full of typos, they are quite clearly exactly how the author presented the work, with just a few minor changes.

*Interviewer*: But they shouldn't be? I mean they should have been changed whereas you would get cross if somebody, one of your people changed a word.

Y: They may not change a word, but just the addition of SGML* coding to a text can radically change it, because you are taking the text and you're trying to interpret what is important about it for the user. So if you go through to where we're keying Arctherus, the Italian Renaissance at the present time and look at the books, the complexity of these Renaissance books with their plates and their colophons and so on and that each of these has got to be accommodated in a computerised form, each

---

*SGML stands for 'Standard Generalized Markup Language'. It is a generic markup language for representing documents which enables information to be separated from its presentation. This makes it possible to present the same information in different ways, for example using a hypertext link between different documents to access particular pieces of information. Hypertext Markup Language (HTML), which is a subset of SGML, makes it possible to have pictures and text on web pages.

illustration's got to have its own particular code number, you've got to deal with captions in a particular way, you've got to deal with the capital letters, it requires real intellectual under-standing of the structure of the particular work.

The conversation continued with Y insisting that Chadwyck-Healey carried out all the functions of a publisher but in a different way. I accepted his statement but said that it did not seem quite the same. He concluded: 'No no I know, don't worry. I've had an inferiority complex about it for, right from the beginning.'

This is an interesting reflection. Y did not appear to be a man with an inferiority complex but by piecing together other conversations it is possible to see a theme. In the early days, many observers believed that the company simply microfilmed to order; in other words, that it did not take a risk in publishing information for sale. Y resented this. He was not a technician providing a service to other creative people. The creativity was embodied in the company. Thus his insistence that adding SGML coding is a creative act tells us how he saw the company and helps to explain the atmosphere in the company and the kind of staff who worked there. It felt like a high-quality publishing house and had a professional and quasi-academic air. The building is modern and decorated with comfortable furnishings and high-quality art. Attention was paid to the physical surroundings and staff treated each other with respect. It felt busy but not frantic.

U, the Sales Manager, confirmed those meanings. The company was essen-tially a publisher which was becoming more systematic. 'We were sort of traditional publisher who just produced for publishing's sake, on [Y's] whim, really. He thought it was a good project. And that instinct has proved to be very successful but now we are looking at ways of perhaps evaluating projects rather more before taking them on.'

And N, the Managing Editor for data conversion, also defined the pub-lishing nature of the company: 'It's to do with the nature of the editorial work that's involved. And typically the editorial work involved in data conversion is, our ideal is to intervene as little as possible in the process of text conversion. I mean, people who come in with editorial skills from other environments expect to be using those skills to change texts in some way or another. What we want to do is to keep the texts unchanged as much as possible, and just put in a layer of information that makes those texts searchable or viewable, so it's actually quite difficult to get people to do that. And it's also those who come in with editorial skills where they are used to either copy editing or annotating texts tend to find it a rather tedious occupation. Because the intellectual input is different, the intellectual input is one of data analysis rather than content analysis. Structure rather than content is important.'

The company is defined here as a specialised publisher with a distinctive style. It is not defined as a conventional academic publisher, but the sense staff have of their work is more than simply reprinting or copying to order.

It had a flat structure. Staff were left to do their own work in a professional atmosphere in which ideas were circulated with the emphasis still on looking for new projects. U commented on this in the context of the establishment of a new department set up to seek out project ideas. He was responding to a suggestion that staff should channel any new ideas through the new department: 'I think now if we had ideas we would perhaps have a more certain route to direct them to. But I think [Y's] idea of the whole company contributing to new projects is a bit fanciful.'

U's comment here shows how the structure of the company had been modified formally so that new ideas could be more easily assessed. The new department was intended to evaluate suggestions and to give staff a route for passing on ideas whether they came out of meetings between senior managers or arose from an observation made by more junior staff. U was cynical about 'the whole company contributing to new projects' but he recognised that communication links had been set up which made this a possibility and enabled better evaluation to take place. The company culture of professionalism and seeking out new projects is well illustrated.

Thus staff believed they would be able to make suggestions and that those suggestions would be taken seriously. Staff had access to decision makers and there was a transparency in the company's procedures.

The basic culture was not democratic, however. There appeared to be a level of trust that staff would take responsibility to circulate ideas or issues which were potentially important. The structure of meetings was relaxed. N said: 'The Board certainly meets monthly. The, there isn't too much information fed down from that meeting. We do have monthly meetings with all of the team leaders stroke project managers in Data Conversion and we started to have, I think, two monthly meetings with the Editorial Department as a whole, but I'm not sure that that's hasn't drifted away.'

In a later conversation, around the time that Lion was becoming established, N commented that there was a risk of drowning in a sea of meetings. He did not want or expect so many meetings. The expectation, therefore, was of a relatively low-key approach with defined but flexible reporting lines. N described staff attitudes to their jobs as equally flexible: 'In the sense that to date, anyway, people have been, job descriptions have been fairly, well job descriptions actually never existed by the way. But the sort of jobs that people have undertaken have been pretty fluid. People have been prepared to move, the skills base we have is fairly non-specific, i.e. we're skilled in doing what Chadwyck-Healey does. And when Chadwyck-Healey does something else tomorrow then we become skilled at that. We have, in Editorial, a fairly flexible work-force.'

The company harnessed the latest information technology. This was a major feature of the company. It did not simply follow the market but was at the forefront, creating the market. However, the perception within the company was that there was not a focus on technology. Y was quite adamant that the latest technology was a tool, usually to get out of a declining sector. Y's statement below perfectly sets the context, as he saw it:

*Interviewer*:   You've always been a company that has used the most up-to-date information technique. You have said to me in the past that that was in a sense only a tool, that was not the objective, but your actions, in a way, almost belie that, because you seem to have always been the first company to use microfiche, the first company to use CD-ROM, and the first company to use the internet extensively. To what extent do you feel it's important for Chadwyck-Healey to be right at the forefront of information technology? Is that something that's important for you?

*Y*:   No it isn't. I mean this may seem strange and from what you've just said but really the delivery technology is not of great interest to us. We always seem to have been using a new technology to escape from a previously unviable situation. And I think this is very, very important to understand, because there obviously are companies which fall, people who fall in love with technologies, who want to be pioneers and I think it would be completely to misunderstand where I come from if you thought that I was one of those people. You know I'm not a Clive Sinclair, I'm not an Apple person, I wish I was. I went into microforms and I was only in kind of insular British eyes would I be considered to be early. Because actually microforms had been around in America certainly since the late thirties and I went into it in 1973. So it was a totally mature market then, and a media, though the microfiche was somewhat new, but even that wasn't that new. It had been around for some time. It was the uses to which they were put that perhaps at that time were very limited. I went into microforms because I had been in reprint publishing before that, and the number of reprints, copies that you could sell of any one reprint had declined to such an extent by 1971/72 that it was no longer, printing was no longer a viable delivery mechanism. And so, but microforms were 'cos you can make a negative and then make money if you only sell ten copies. And that suited the academic environment perfectly. By the mid-eighties microforms for the publisher, let alone the user, had become a pretty boring and limited publishing medium. When the CD-ROM came along it looked enormously attractive in being able to do so much that the microfilm had never been able to do, but equally it was glamorous. It was glamorous to both the publisher and to the user. And the economics weren't that different. You still made a master which wasn't incredibly expensive to make, and then made relatively small numbers of copies. And then here we are a year later, or a few years later but a year ago, with our CD-ROM sales declining because libraries are saying 'We do not want CD-ROM', so we say to them 'What do you want?' 'We want it delivered on-line.' 'Ah, thank God for that. There's another way we can do it, which is actually even cheaper in some ways.' But then what is different between, I think what we're doing now and what we've ever done before is that actually the internet – it could be any on-line system, but the internet particularly lends itself to this – offers truly new and creative ways of publishing, ways of linking information together. Which set it apart from anything else I've certainly ever been involved with.

The company's prime business was the publishing idea, but Y was fascinated by the possibilities created by electronic media. He claimed he was not someone who had fallen in love with the technology, and the main activities in which the company was involved confirm this. Staff were busy coding complex literary texts or discussing demanding literary works with academics. When asked what features constituted a 'Chadwyck-Healey' person, Y put high up on his list a need not to be overwhelmed by or uninterested in academic ideas. Other staff confirmed their orientation to these academic or publishing ideals. But there was nevertheless an underlying interest and excitement about carrying out traditional publishing activities in a complex technological way, particularly when, as with CD-ROM and even more with on-line delivery, such methods opened up possibilities unavailable to the traditional forms, for example using hypertext links between different texts and integrating textual analysis into the text itself.

The culture at Chadwyck-Healey was therefore a professional one based around the conventions of academic publishing but with an unconventional (in publishing terms) concentration and enthusiasm for electronic media. Relationships were flat, staff had a degree of autonomy, though that had not always been the case, and there was an expectation that relationships and job functions must be viewed as flexible in the light of potential changes in the company's market. Staff believed that their ideas and suggestions would be taken seriously. Underlying this, the focus of the company was interpreted as the search for new ideas and new projects. Y remained at the heart of the company as the primary source of new ideas and its respected Chairman. He had moved away from a deeply autocratic style which he operated in the early years of the company, though he continued to retain elements of that approach. Routines were reproduced and modified in this culture with the underlying focus on the search for new product areas infiltrating all that was done.

# Agency, structure and routines

## Agency

The primary agent in the company was Y. His influence was enormous. He had responded to the environment in which the company was placed and had used his influence in different ways, first as the prime mover and more recently as a relatively hands-off Chairman. Of the four companies studied in this project, Y's centrality as a key agent is the strongest. D, the founder of Unipalm-Pipex, while both more involved in the day-to-day operation and more aggressively influential in the detail of Unipalm-Pipex, did not have Y's (relatively) unchallenged control. Y's influence came from the same source as D's – he was the company founder and was highly knowledgeable and skilled at spotting market changes – but the company had a much longer history and Y clearly had the respect of his staff so that his influence could be brought to bear in barely perceptible ways.

The company's style of working closely matched the configuration Mintzberg describes as professional (Mintzberg 1978): individuals are highly skilled and work largely independently and autonomously. Project-based firms are typical of such configurations. The staff interviewed were, in that sense, important agents in their own fields. Each brought a particular expertise which was respected by other staff.

## Structures and routines

Internally the company structure reflected its professional approach. It was open and flat with autonomous departments. Externally Chadwyck-Healey saw itself as operating in a particular niche. It concentrated on academic texts which did not have mass market appeal. Y and U (the UK and Eire Sales Manager) both commented at different times that major corporations who might potentially become involved in the niche were unlikely to do so. Microsoft, for example, would not see Chadwyck-Healey's success in delivering academic texts through the internet as threatening to its business, or probably as an area ripe for profitable take-over, simply because this form of academic publishing is so specialised. The company did not operate in an atmosphere of threat from other players in the market, therefore. There was uncertainty and threat in questions about whether it could earn sufficient revenue from a specialised niche, and what it must do to maintain the interest of its customers, but relatively little concern over other companies supplying similar products.

The company operated in its market place largely through networks. In this respect its routines had similarities with the National Extension College, which is also involved in publishing, and there is probably an element of industry recipe implicit in this style of working. In its production activities, too, networks and close, relatively informal relationships had become increasingly noticeable. The networks were tightly intermeshed and not just a loose collection of relationships. N described the way in which his complex conversion processes were managed and how decisions taken in-house about how to define data – to tag them – were now shifting to suppliers who keyed the data into electronic databases: 'We would define that, the way in which the data's to be tagged. But there's the initial editorial work of defining the tag set and doing the data analysis, but then there's the sort of ongoing editorial process of every time you get an actual real instance of the data and compare it against your ideal specification you find the two never quite marry up. So there are some editorial decisions to be made on the fly, as it were. Is this an argument, is it an epigraph, is it a note or whatever. Increasingly we are relying upon the keyer to make those sort of decisions, although we still guide them, we have an editorial team, that's essentially what the data conversion team do is mark up text to guide the keyers. But that's where we are currently experiencing the sort of growth of the symbiotic relationship with us as publishers and keyers as data providers. They can do more, take on more and more of that.'

N was concerned that the company should not become too reliant on suppliers. He recognised the down-side of the 'growth of the symbiotic relationship'. The keying in of texts – transforming them from written to electronic form – was substantially subcontracted. Chadwyck-Healey staff carried out complex editorial functions. They stated the precise way in which different conventions in the relevant literature were to be defined and therefore the way in which the structure of the piece would be interpreted – 'defining the tag set'. But the basic keying was carried out in Asia, and increasingly editorial functions were carried on there too. The relationship with these supplier companies was relatively informal (there were no formal contracts beyond single requests for purchase of services) even though the Asian companies depended heavily upon them, and in at least one case they depended on the Asian company. Initially N used agents in the UK to maintain contact with the supplier companies. Later he developed his own relationships with them, and increasingly day-to-day negotiations were carried on by his staff with their counterparts in the companies concerned, through telephone, fax and e-mail. The use of such networks enabled the company to maintain a strategic fit with its suppliers – the styles of working in the companies were broadly aligned – and was a source of flexibility when products changed. N was concerned to ensure that the company did not become heavily dependent on one supplier but his relationships with the companies were relaxed and personal. It was clear that the boundaries between suppliers and the company were becoming blurred as the suppliers took on work which had previously been carried out in-house and relationships became more relaxed and interconnected. Thus operational routines had evolved and felt comfortable. These practices had implications for strategic routines, as will be explored later.

Similar findings have been reported in other organisations which operate with a networked international division of labour of this kind (Ngwenyama 1998). The relationships tend to become more informal and more trusting. Such developments were not surprising, therefore, and arose partly through the richer communication possible with the new electronic communication technologies.

Networks also formed vital parts of the company's sales operation. U and his colleagues maintained relationships with major academic libraries and individual librarians but they also developed relationships with individual academics. The academics, as library users, needed to be convinced that Chadwyck-Healey's electronic products met their needs and the libraries needed to be convinced that they were worth purchasing. There were many people involved but this was not a mass market. Relationships were developed at conferences and through e-mail networks as well as by personal contact.

The relationship is one of 'voice', to use Hirschman's term, not 'exit' (Hirschman 1970). The simple distinction between exit and voice can be helpful. In a dynamic context voice becomes an increasingly important feature, as firms listen to their customers and try to retain their custom. Voice is the exercise of complaints or comments by customers, who may well wish to remain with the company but want it to perform differently. This is particularly significant in sectors where repeat or continuing business forms a major consideration. Exit, the conventional market behaviour (in neo-classical economics) of taking business elsewhere, may be an undesirable outcome for both buyers and sellers. In distinguishing between these two forms of market operation the socially constructed nature of markets is demonstrated. In practice, as Hirschman points out, markets are different. The market for Nigerian railways (to use Hirschman's example) is different from the market for schools. Markets are even defined differently in legal terms. The obligations of buyers and sellers in the housing market are legally different from the obligations of buyers and sellers of vegetables. In interpreting markets and developing routines with respect to suppliers and customers, the possibilities for exit and voice will form part of firms' strategies. A firm may use different strategies in different parts of its operation. The possibility of exercising voice – perhaps through electronic communications – may be one of the considerations for the firm in defining the permeability of its boundaries with other firms and the extent to which it attempts to police those boundaries. There is an example of that here: the users of Chadwyck-Healey's products frequently discussed their needs with the company and commented when things were not as they would wish. They did not transfer immediately to alternative suppliers. This was an approach which the company encouraged. It had parallels with the approach to production. Relationships were personal and flexible – 'voice' – not simple market exchanges.

As an example of the evolving nature of this kind of network, U talked of the need to visit customers largely out of courtesy. He then made an interesting comment on the changes e-mail use had brought: 'E-mail and internet are becoming more important for us. I mean I tend to do a lot of my communication with customers now by e-mail. I think actually that's led to an improvement in communications from my point of view. 'Cos academics and librarians are very happy to use e-mail. Particularly at this time of year, you know the first few weeks of a new term the library in the university is pretty chaotic. It's not really a time when you can phone the librarian up and expect a relaxed conversation.'

He continued later: 'I think that's a very useful way of doing it. And also another example would be I read about it, the announcement of a one-day seminar on a particular date, from a librarian in a university in the UK, and he mentioned in the agenda for the seminar that he might be looking at some sources of information in a particular subject field and mentioned one of our products, so I e-mailed him and asked him if he'd like to have a copy of that, he was very grateful. . . . I was attempting to, not pressurise him into buying something, but offer something which he thought might be beneficial to his seminar, and that we would benefit from because he would be promoting our product.'

Thus U's use of networks and the way in which he has used e-mail to extend his opportunities reflect a developing set of personal relationships in which the particular requirements of the customer are taken seriously. Similarly at a strategic level, visiting contacts in libraries throughout the world was an important part of strategy formation, as well as attendance at conferences and the maintenance of other networks. The company has small subsidiaries in Europe and North America which also formed part of the information-generating network.

A primary routine at both strategic and operational levels is thus the use of networks and the development of personal relationships in all the company's activities. Other routines interacted with this broad orientation, however. While networking was a preferred and common pattern of working, and there was a tendency in the company to adopt that approach, it was sometimes challenged. The Lion story goes on to indicate the interactions of routines at Chadwyck-Healey. The development of routines can best be illustrated by considering a particular set of manifestations of them. The Lion story explores a period of significant change in the company's history. Management of this change revealed many of the more implicit aspects of the company's behaviour and its examination enables us to acquire a clearer and more subtle understanding of routines at Chadwyck-Healey.

The Lion story reveals that the use of networks was not the only element of the strategic routine. It was a preferred orientation and was closely aligned with the operational practices of the company. The existence of such practices was of substantial assistance in developing Lion. But in incorporating change, networks and their associated practices did not appear to be sufficient.

## The Lion story

During the fieldwork, Chadwyck-Healey moved from being a publisher providing its product on CD-ROMs to one which delivered on-line. It still produced CDs but the future of the company was seen to lie in on-line delivery. This was a major shift. It involved accommodations with, and developments in, company routines and illustrates the way in which the company searched (in the Nelson and Winter (1982) sense) for effective routines, and learned from its environment.

The idea was first mooted as a 'virtual library' in the middle of 1995. It appeared fully fledged as 'Literature on Line' (Lion) in December 1996. It was first brought to my attention in the early autumn of 1995. N explained that the basic idea had arisen from arrangements that had been in place for some time with a number of American universities. Basically, they had not purchased Chadwyck-Healey's databases on CD-ROM but had bought the data and put it onto their own electronic servers so that their users could access the material directly on personal computers throughout the university. Chadwyck-Healey therefore began to consider seriously the idea of providing client-server versions of their products. This was tied in with the idea of creating what N called an 'old cliché', the virtual library of data: once there is a client-server system in place for databases, then the question of searching simultaneously across more than one database is raised. Chadwyck-Healey were therefore looking at ways of harmonising data structures so that people would be able to search across various databases at once, and at ways of tying that in to other information sources. I asked where the virtual library idea came from:

N:             [Y].
*Interviewer*: Just a bright idea one day or?
N:             Yes. Exactly was.
*Interviewer*: What's your view on it?
N:             I have to say I think he wanted to run before we could walk a
               bit. The virtual library sort of pre-dated the file server versions
               of the databases. My feeling was that we were developed, we
               were moving in that, or we would have moved in that direction
               anyway, [Y] has simply wanted to accelerate the process and
               still wants to accelerate the process. And is pushing us all to
               run faster as a result. I'm concerned that we might have gone
               too fast but.
*Interviewer*: The idea's OK?
N:             The idea's excellent. [Y's] concept of what is achievable, I think
               is a bit optimistic but then that's what he's there for. And we
               may well end up achieving 90 per cent of what he thinks we
               can achieve.

Having been alerted to the 'virtual library', I picked it up in a conversation I had later with U. At this time the idea was one of many I had heard circulating in the company. It had some push behind it but I did not see it, at that time, as something enormously significant.

U's view was that most users had become less concerned with the medium of delivery, and that they were considering the use of databases obtained via some sort of on-line link. This was driven partly by the increasing use in libraries of the internet as an information source. He was therefore confirming the viability of the project idea but knew very little about it. U had a broad grasp of the project, but at that time, having spent some time discussing matters relating to the Editorial Department, I knew more of the detail than he did. The project was clearly not at the forefront of staff minds.

A more pressing issue for the Sales Department was the sale of the Chadwyck-Healey databases to a national consortium in the UK (CHEST) which in turn made them available to all UK universities. This had relevance for Lion, as will become clear, but at that time was viewed as an entirely separate issue. In the autumn of 1995, there seemed to be little connection between the two projects. The CHEST deal concerned U since it changed his market place substantially. It meant that all universities in the consortium now had access to Chadwyck-Healey's products simply through their membership of the CHEST consortium. Since a major part of U's market lay with the universities, this had consequences for the way in which he carried out his job.

In terms of explaining the company's move to Lion, the two developments are connected. CHEST changed the company's perception of part of its market. Selling databases to consortia, for the members of the consortia to use as they saw fit, alongside the American experience of client servers, was an indication that the market for individual sales of expensive CD-ROMs might not be robust. There were implications for the kinds of approaches which could be made to the market and the kind of expectations they would have of the company. As U put it: 'So basically what we've done is we've just sold, instead of selling the data fifty or a hundred times we've sold it once.'

Within that short phrase there are many implications for the company. It began to see its market differently. Its interpretation of its obligations to its customers had changed. It sold products on the basis of their content and value to the customer. On this occasion it then later negotiated very different sales arrangements which were a much better deal for the customer than those initially constructed. CHEST was not easy for the company. Universities' expectations that they would always be treated well were challenged, not because they were treated badly, but because they paid out large sums for something which would have seemed unnecessary had they had more information about possible future relationships. At a different time, I saw some of the correspondence from the Vice-Chancellor of one university to Y which made it clear that this was seen as a significant issue of trust and fairness. The fact that the Vice-Chancellor had chosen to write confirms the 'voice' nature of Chadwyck-Healey's market place. Sales staff were not accused of double-dealing but their customers became more wary of the precise terms of sale. The networks which had been carefully nurtured were threatened by this shift of approach and I understand Chadwyck-Healey did make some compensation payments, though there was no legal obligation to do so, in order to maintain good relationships with customers.

Thus although the CHEST arrangements were largely separate from what eventually became Lion, the way in which they began to open up different interpretations of the market help to account for the later developments. Agents' perceptions changed and the structures of the market shifted.

In the meantime, Y was developing the idea of the virtual library. A new Editorial Director had recently been appointed, partly to bring more creative ideas into the company, and he and Y had been spending a lot of time acquiring rights to other works and setting things up. I asked Y where the idea for Lion came from: 'Interestingly it did just come as a kind of, I suddenly had a vision of the whole thing sitting on a plane going to Australia in April. My wife and I were going for a three-week holiday. I'd taken a few articles relating to work with me to read on the plane. We'd worked very hard, worked very hard to get ready for this trip, and I think in a strange way you get onto the plane, you're cocooned for twenty-four hours or whatever it is, twenty hours, and you suddenly realise you can relax. And your mind begins to work in a different way, and I read something in one of these articles which just triggered off a complete new line of thought about how the fact that we had made no attempt to link all our databases together, and if we linked them together why shouldn't they be linked to other people's databases? And quite honestly the whole thing certainly evolved in my mind in five minutes. It was just, it was suddenly so obvious I couldn't think why we'd never thought about it before.'

This is an extreme example of the way in which Y came up with ideas for the company, but it is not untypical of the kind of behaviour which those interviewed expected. His excitement about the idea related to the opportunities presented by building links into databases. At that time his thoughts were less developed than they were subsequently but one can see the foundations of the new project.

We should now travel on in time to December 1996 and look back at the trajectory of the project and what had to be done to make it work. In mid-1995 it was clear that company approaches were shifting as the company learned about its new environment. Six months later, in January 1996, the project had been accepted and understood by staff but it had not been incorporated into day-to-day activity. Y described his realisation of this:

Y:      But I came back from Christmas break, had a Board meeting and we were discussing, by this time we were discussing our production plan for the coming year in enormous detail, down to the tiniest, most insignificant little CD-ROM. The one thing missing from the whole plan was our literature web-site. And I realised that it had got, it had really been overlooked because it didn't fit a slot. It wasn't being looked on as a product . . . I realised that we weren't, there was something we weren't doing right, 'cos here we had great idea, we'd already decided months before that it was probably the most important thing the company should be doing, and yet we weren't doing it. And to me that's a fascinating management challenge. What happens to ideas? Why do well-intentioned, hard-working people fail to do what is quite clear that they should be doing? Well one of the ways to break these log jams is for somebody to lose their temper, which is virtually what happened on this occasion.

*Interviewer*: Who lost their temper?

Y: I did. And I [one Director] said that she couldn't do it, no, she thought that she could possibly do it by spring '97. This is January '96. And I said that that just felt too long to me, that I felt we were going to be overtaken by events. And that the first week in December '96 should be the target date. I went off to Russia the next day, for a few days, and I came back a week later or so. . . . [The Director of Sales] had really pulled everybody together and had got a plan going. He bought some very good planning software, I can't remember what it's called, and we delivered what is, I believe, one of the most sophisticated uses of the web and it's certainly user friendly, works beautifully, incredibly solid, incredibly large and we delivered it this week, last week, first week of December. So that was a real triumph, actually. But it was because everybody got together, became totally focused, drew up a, you know, a work-flow, a chart that really was done properly, and it recognised all the key steps and then stuck to it. And, you know, we do have the production depth here to actually know how to do the thing anyway.

A good deal happened between Y losing his temper and the delivery of Lion. The directors had suggested setting up a committee to assess the needs of Lion. But no one person was able to take responsibility. At that time the company did not have a Managing Director. There were four directors, each responsible for a different part of the company, reflecting the professional model to which they implicitly ascribed, and a Chairman (Y) who did not have a line management role and was, in his own words, 'in and out, disappear for weeks at a time, not be particularly close to day-to-day work and development, kind of throwing ideas into the pot'. The previous Managing Director, who had left some time before, had not been replaced but now there was a sense of a vacuum. Y had a sense of dissatisfaction and impatience that something was missing. He felt the company had got bogged down and things were not advancing in the way that they should. The company had been having weak trading results which also contributed to the dissatisfaction. In February another very stormy and difficult meeting took place in San Antonio, Texas, one of the regular three-monthly Heads of Companies meetings when the heads of all parts of the Chadwyck-Healey group get together. The meeting discussed the weakening sales position in Europe and problems with the American company as well as new product strategies. Y claimed he had enjoyed all three days of the meeting but the next day he drove to Houston with the Director of Sales and was told that by the end of the third day all the directors were ready to resign. Whether all the directors felt so strongly is not clear but it is undoubtedly the case that there was a great deal of unrest. The outcome was that in April the Director of Sales was made Managing Director.

Here, then, we can observe plenty of disruption. The company was open in its relationships, however, and there were opportunities to discuss, argue and learn from the dissatisfaction. Matters did not become undiscussable (Argyris 1976), so that the Director of Sales, for example, was able to raise his perceptions of the difficulties with Y. In this case new structures were created from self-conscious reflections on previous structures by senior staff and as a result of the behaviour exhibited by those staff. This was both enabling behaviour – the Director of Sales' pulling together of the team – and constraining behaviour – the reluctance of others to move quickly. The search for new structures produced modified routines, for example in redefining the idea of a project, so that Lion was seen as, in some senses, equivalent to existing CD-ROM projects. The search was driven by discomfort and dissatisfaction and the new structures were created by the recursive interaction of existing structures (within the company and market) and agents not by the imposition of a fixed plan or decision to change. The trust and confidence of all staff was an important part of the company's ability to flex in this way.

At a more detailed level there were a number of significant changes required to facilitate the shift to on-line distribution. Y was proud of the company's financial conservatism but the move to Lion had to be financed, and existing conservative operational routines were insufficient:

Y:          We've probably borrowed already just about as much as we can, because the other big thing in '96 we did a complete refinancing with Coutts, but I would consider selling equity at this stage.

*Interviewer*: You would?

Y:          Yes. Which I certainly never have been prepared to do before. But because we have other interests in other companies in the US it's more likely that we would sell those interests to raise money. And in fact some of that is possibly quite imminent.

My surprise at Y's acceptance of the possibility of outside equity finance is clear. For Y, Lion represented an enormously important change. He described it as a huge change for the company. The new willingness to consider equity finance gave clear signals about the size of the task. On the financial side of the company, F (the Chief Accountant) recognised that moving products to delivery by the internet was a risky venture and that companies which were not privately owned might not be prepared to take that risk. He showed an element of pride in the company in making such an assessment. He was not concerned about the refinancing of the company. Given the company's growth and profitability, he was surprised it had not been done before, and while it was clearly a shift in company style, the move to properly secured loan finance did not seem extravagant. It was a continuation of existing policies. The existing conservative financial routine was being translated for the needs of the larger company that Chadwyck-Healey had become. A move to external equity finance would have been a shock, and would have the potential to transform the company, but during the fieldwork this did not happen.

Financially, then, while significant shifts had been contemplated, the company was moving along a track which matched previous behaviour. The financial operational routine was being modified and new possibilities were emerging, but there had not been disruption and wholesale change.

In sales and production the move to on-line delivery had resulted in some profound changes. In both areas staff had become uncertain about their jobs and had sensed that communications within the company had grown weaker. The stormy nature of Board meetings and Heads of Companies' meetings had not filtered down but a much higher than normal number of staff had left. The staff interviewed explained the higher than usual loss of staff as a function of the size of the company or as age-related, and because pay and conditions had dropped below industry standards. They believed the problem had been largely resolved by the introduction of a more formal personnel function. The potential need for a clearer focus for personnel policies and practices as the company grew had been commented on earlier by a number of respondents. It is typical of the company that it moved relatively quickly to put such features in place and this matches the pattern of other changes: the promotion of the Director of Sales to Managing Director, the continuation of conservative financial practices in new contexts, and the greater formalisation of personnel functions – evolutionary changes not quantum leaps.

Externally, the developing relationship with suppliers – the strategic fit of routines between them and the company – enabled elements of dislocation to be managed without major disruption. N was aware that activities like Lion would require less use of some data conversion processes. For example, the keying companies were potentially required less because more data would be transferred from existing CD-ROM products which were already coded appropriately. The informal relationships, rather than contractual commitments, developed between N's section and the companies enabled him to manage changes through the trust he had developed in his networks; he had already negotiated reductions in supply from one of the supplying companies.

The company was remarkably non-political in comparison with many other institutions. While there were grumbles and complaints about, for example, bonus payments and salary levels, there was no evidence of significant in-fighting or manoeuvring. But during this period of change more dissatisfaction with processes and direction was evident than before, as shown by, for example, the suggestion that a number of directors were ready to resign and the higher number of staff leaving. It was clear that what had previously been taken for granted about the way the company operated could no longer be completely assumed and this created uncertainty.

In summary, the company was operating with a broad orientation, perhaps reflecting a recipe for specialised publishing companies, which emphasised networks and personal relationships. This worked at strategic and operational levels and was an important part of relationships with suppliers and customers.

The move to Lion matched the company mantra, 'You're only as good as your next project'. But it did not fit well initially because the idea of Lion did not match the company's view of what constituted a project. Y lost his temper – Action Man with a polite face – in order to force through this change in meanings. He had proposed a major change in product mix, which was part of normal practice, but it had not initially been interpreted as a project by others in the company. It did not match perceptions of projects as, for example, smaller-scale CD-ROM ideas. (But it is worth noting that CD projects at Chadwyck-Healey were large scale by normal standards. Chadwyck-Healey CDs cost thousands of pounds to purchase, but the new project was orders of magnitude greater than that.) Y acted out some elements of D's Action Man. He forced through change at a pace greater than that initially seen as possible by his colleagues. This created some disruption and uncertainty within the company.

While I was unable to observe Y's version of Action Man, it is inconceivable that it included the explicitly aggressive and abrasive behaviour found at Unipalm-Pipex. But the underlying motivation was very similar: move fast and take irrevocable steps. In forcing change and pushing senior staff to modify the company's taken-for-granted way of working, Y was following a strategic routine based on the widely accepted interpretation of the company's world that 'you're only as good as your next project'. That phrase is not a routine. It is a shorthand way of capturing the company's shared meaning of its external environment. The routine follows from an acceptance of that interpretation. What was interesting here was that Y enthusiastically took a leap of faith and demanded that his colleagues redefined an important element of the company's shared meanings: the nature of a project.

Externally there were many similarities with Unipalm-Pipex: the product was new and untested and created a new market. But at Chadwyck-Healey this did not generate aggressive, competitive behaviour.

The company's strategic and operational complementarity in its routine behaviours, with each element enabling and feeding into the others, was important. For example, N revealed the way in which his informal relationships with suppliers were enabling in permitting strategic changes to take place. Long-term contractual agreements would have at least slowed down changes and might have made it impossible for the company to respond to the opportunities it foresaw. His networks, and those of his staff, had not been developed as a result of a deliberate company policy but had been set up in the course of working with the supplier companies and were a feeding through of the taken-for-granted way in which Chadwyck-Healey carried out its business.

The strategic routine at Chadwyck-Healey thus had a number of components. It operated through networks which were aligned closely with operational routines. It evolved. For example, the move to on-line distribution progressed through schemes such as CHEST. It was also based around a focus on the next project which itself had formal and informal components. The main source of new ideas was Y. N said, '[Y's] concept of what is achievable, I think is a bit optimistic but then that's what he's there for' and this informal role was recognised by all respondents, including Y. It was partly the recognition of the reliance on Y that prompted the setting up of a new department to consider new projects. U believed that '[Y's] idea of the whole company contributing to new projects is a bit fanciful' but the fact that formal structures were set up to try to harness new ideas captures the emphasis in the company on this aspect of their work. (At the end of the fieldwork the new department was reportedly less successful than had been hoped but that does not diminish the emphasis its creation gives to that aspect of strategic behaviour.) Finally, but only infrequently, Y adopted Action Man tactics to force change when he believed it was being unreasonably resisted.

## Technology as a constitutive element

The technology never featured as a major inspiration or company focus at Chadwyck-Healey. The company was dealing with the most up-to-date technology but, as quoted above, Y claimed, 'the delivery technology is not of great interest to us. We always seem to have been using a new technology to escape from a previously unviable situation.' He clearly enjoyed using modern methods. When challenged on the company's relatively late acquisition of an e-mail system, he pointed out more than once that he had been one of the intrepid users of 'Telecom Gold' when the term 'internet' had not been invented.

Electronic mail was implicated in changes in approaches to external agents, as has already been discussed, and in internal structures. U said: 'I think it is actually this has been helped a lot by the internal e-mail. Because since that went live information has been provided much more readily. I think there was a tendency, at one stage during the company's, in my time at the company, where information wasn't really filtering through very well and whereas, I mean this goes back to your sort of sales meetings and product information meetings which you might expect to have on a fairly regular basis in a publishing house, but we weren't really getting the information we needed to sell these titles. Now things are improved greatly because, obviously, you get an e-mail message, anything of any interest is posted on a bulletin board about, which is available for each project, so information is much more readily available.'

E-mail fitted well with the company's flat structure and enhanced opportunities to maintain professional autonomy and share information. But the adoption of e-mail was relatively late. In an early interview, N commented that the company was in many respects technologically backward:

N:　　　　If you're asking how important are they [new technologies] internally? That's? That I think has still to be proved, yet. We, despite the fact that we are a relatively high-tech operation for a publisher I think internally we operate in a fairly low-tech way.

*Interviewer*:　By which you mean no e-mail?

N:　　　　By which I mean no e-mail and no connection to networks, not even a PC necessarily on everyone's desk.

E-mail was widely and readily adopted (during the time of the fieldwork) and improved communication flows. The company was conservative about taking up some of the opportunities provided by e-mail, however. For example, there were no significant moves to homeworking even though this would have been compatible with the method of working and relative autonomy of staff in Sales particularly. Some staff had suggested homeworking and the use of appropriate technology as a desirable option but it had not been taken further.

The company did not focus on technology or relish it. N's concentration in his work, as quoted earlier, was that 'the intellectual input is one of data analysis rather than content analysis. Structure rather than content is important.' He was concerned with the complexity of the written word, not with the technology which enabled him to reveal it in certain ways. Similarly U used e-mail to maintain networks which he might otherwise have managed in other ways. He found opportunities which arose only through the existence of e-mail so that he could get closer to his clients and make unexpected offers to them. But the technology itself was of no interest to him.

This does not mean that the technology was unimportant, quite the contrary, but staff at Chadwyck-Healey used technology in different ways, in the manner in which pencils, calculators and typewriters were seen as tools in earlier decades. Unlike pens and pencils, the technology in this case, however, was implicated in all that was undertaken. Much as at Unipalm-Pipex, if there had been no electronic technologies there would be no

Chadwyck-Healey, but at Chadwyck-Healey it was a medium rather than a *raison d'être*.

Chadwyck-Healey is a publisher which also happens to be a high-tech company. It could, with its current product structure and orientation, equally be a high-tech company which happened to be a publisher. The history and culture of the company make it the former. Its world is interpreted by staff using a broadly publishing perspective. Operational routines and products, however, are intensely technological. In some senses the technology is more deeply constitutive of the company than at Unipalm-Pipex because it is literally taken for granted by the vast majority of staff. It is not of interest but is assumed. And yet without the technology the company's internal and external relationships and products could not have become what they are. As at Unipalm-Pipex, the social and technological are practically inseparable.

## Unipalm-Pipex and Chadwyck-Healey: contrast and comparison

Operational and strategic routines at Chadwyck-Healey were centred on well-developed networks, which formed vital parts of the company's operational style. These were set in a context of continually searching for new project ideas which gave the company a thoughtful, perceptive and self-reflexive appearance, and of the harnessing of the latest information technology within the world of information provision, i.e. publishing. Such self-reflexivity resulted in the adoption of new technologies largely in response to perceived difficulties in existing media. The technology itself was used conservatively; for example, as mentioned above, there was a resistance to homeworking. Financial conservatism was also maintained even during the unprecedented changes made to develop Lion.

The company interpreted itself as a high-quality publisher and operated in a professional publishing style. Y occasionally exercised his autocratic powers and adopted many features of D's Action Man in trying to create a new market.

Both Unipalm-Pipex and Chadwyck-Healey were prime movers, some would argue *the* prime movers, in inventing (at least parts of) their markets and in doing so faced major changes in company orientation. Both operated in conditions of uncertainty in circumstances which were changing rapidly. A key feature for both companies was that they were exploiting the cutting edge of electronic technologies, both in fact dealing with the opportunities provided by the internet. In neither case was the product provided by the company conceptually complex: Unipalm-Pipex connected consumers to a network of cables down which they could send messages; Chadwyck-Healey provided academic texts in such a way that links between them could be traced. Both companies had strong founders. Both adopted the strategic practice: move fast and take irrevocable steps. Why, then, did they develop in such different ways? What lies behind their different trajectories?

The cultures of the two companies were different. The aggressive, macho, hard-edged interpretations of Unipalm-Pipex differed markedly from the meanings which arose from Chadwyck-Healey's professional, understated and academic culture. The cultures themselves partly derived their meanings from the different industry recipes. At Unipalm-Pipex the recipe, to the extent that it can be observed, appeared to match a non-conformist, arrogant, slightly wacky representation of the internet. In contrast, Chadwyck-Healey had the jealously guarded, professional, high-quality publisher recipe. Thus the worlds which the companies used as their primary reference points were different in approach and history.

Company organisational structures were different. Unipalm began as a group of companies designed to spin off from each other. They were different in orientation and financed by the cash cow provided by the TCP/IP franchise. There was a multi-cultural dimension at Unipalm which became increasingly bitter as Unipalm and Pipex became more established. Unipalm-Pipex was also a public limited company. Chadwyck-Healey, a private limited company, in contrast, has moved from product area to product area, retaining elements of microforms as it moved into CD-ROMS and retaining much of its CD-ROM work as it moved into on-line distribution. The nature of the product, as against its mode of delivery, has remained broadly constant throughout the life of the company. This has produced a focused company structure in which trust had developed and a set of taken-for-granted ways of doing things – routines – at all levels in the company had become established. The histories of the companies have played a part in the structures they have developed. Chadwyck-Healey has a relatively long history and developed relatively slowly until very recently. It has been able to build a structure and way of doing things in a relatively slow and evolutionary way. The contrast at Unipalm-Pipex is dramatic. It had a short and turbulent history. It faced more external change than Chadwyck-Healey, though some of this was driven by the company itself and was not independent of the other explanatory factors. A routine to force change emerged at Unipalm-Pipex but for most of the company's history it was inchoate. Chadwyck-Healey established networks and other professional routines which meshed well together and saw it through significant transformations.

Agency played a part recursively and discursively in these transformations. Both companies had strong-minded, dominant strategic agents, and both had energetic groups of staff who aligned themselves with company goals, to the extent that such goals could be discerned. At Unipalm-Pipex agents worked independently and sometimes in opposition to each other. They expressed themselves, at times, aggressively. Relationships appeared highly charged. Chadwyck-Healey, in contrast, gave autonomy to staff but within a clearer sense of the needs and responsibilities of different staff groups. There were greater elements of trust and no discernible back-stabbing. It was quite clear from Y's own admission and from reflection on the earlier history of the company that he could, and did occasionally, operate in his own version of

Action Man mode. He reflected on this in relation to the establishment of Lion: 'And to me that's a fascinating management challenge. What happens to ideas? Why do well-intentioned, hard-working people fail to do what is quite clear that they should be doing? Well one of the ways to break these log jams is for somebody to lose their temper, which is virtually what happened on this occasion.' D made similar comments but seemed unable to stand back so much from his actions. In the Unipalm-Pipex case this approach became a dominant one, whereas at Chadwyck-Healey it was less central to the strategic options available.

The evolution of anything is less dramatic when the rate of change is slower and, clearly, the rate of change at the two companies differed. Change could have been sharper at Chadwyck-Healey if the company had not chosen to maintain itself in a relatively narrow niche. Alternatively, Unipalm-Pipex could perhaps have selected a niche market and thus have brought greater calm into its proceedings, though the changing nature of that market during the period of the fieldwork was so great that the niche of, for example, corporate connections which Unipalm-Pipex chose initially was also itself changing rapidly. The sheer rate of change was an influential factor but the companies themselves were partly responsible for the changes affecting them. The decisions to operate in the external world, in the manner in which each company chose, were themselves dependent on the prevailing structural circumstances and agents' activities. History, culture and routine behaviour were implicated in the rate of change faced by each company. Unipalm-Pipex then found itself operating in an environment in which other companies were also active, whereas Chadwyck-Healey defined itself as a specialist company with a unique product. Chadwyck-Healey faced competition from other providers of academic texts but the possibility for substitution between products was much more constrained than in the case of Unipalm-Pipex. It was able therefore to gain more day-to-day control over its environment but unable to grow at the rate of Unipalm-Pipex. The rate of change emerged from factors operating in the different circumstances of the two companies but it was not something which arose entirely independently of their behaviour.

'Technology as artefact' (to use Orlikowski's (1995) distinction) was remarkably similar in the two companies. Both companies eventually were centrally concerned with the internet. Technology in use, however, was very different. Unipalm-Pipex was staffed by technologists who defined themselves in terms of the technology. Chadwyck-Healey was staffed by professionals oriented towards publishing who saw the technology as incidental. Technology was clearly implicated in change at the two companies. In some respects the practical use of the technology at Chadwyck-Healey was more sophisticated than at Unipalm-Pipex but it was a much less important defining characteristic of the company.

Through the complementarity of its routine behaviours and its self-reflexiveness, Chadwyck-Healey was able to search its environment successfully and to learn from it. Change was (partly) managed rather than dealt with as it happened. Conservative financial policies, professional networks and an excitement with the potential of new technologies were all implicated in change in an interconnected, recursive way. A key component is the cultural emphasis on the next project rather than on making as much as possible from existing areas. Unipalm-Pipex created its own market by driving hard and Chadwyck-Healey was very similar. Unipalm-Pipex's market was under threat in an obvious and omnipresent way. The changes faced by Chadwyck-Healey were less threatening.

Thus, in making a comparison between the two companies, we can see how differences in culture, structure and agency, which underlay the practices of the companies, produced different responses in broadly similar circumstances. The consideration of routines and the factors behind them is thus confirmed as a valuable approach in understanding firms' behaviour. Both companies were able to incorporate change into their routine behaviour.

# 8   Concluding remarks

## Introduction

In 1993 James March and Herbert Simon reflected on their classic text, *Organisations*, written more than a third of a century earlier, and suggested that most of what was in the book remained useful. If they were rewriting it they would change relatively little, but there would be four broad themes that would be somewhat different:

> (1) we would give more attention to empirical observations as opposed to theoretical speculations; (2) we would place relatively less emphasis on analytically rational, as opposed to rule based, action; (3) we would less often take the premises of decisions as given exogenously; (4) we would accord a greater role to the historical, social, and interpretive contexts of organisations.
>
> (March and Simon 1993, p. 302)

As a result of their reflections, March and Simon suggest that 'much of the behaviour we observe in organizations is "intuitive" in the sense that it occurs immediately upon recognition of a situation [and] . . . much of the intelligence we observe in organizational action comes not from explicit analysis but from rules' (March and Simon 1993, pp. 309–10).

The case studies set out in the earlier chapters recognise March and Simon's 1993 insights and themes. The categories which are discussed and developed provide explanations for the behaviour of four firms and provide new building blocks for the understanding of other firms in different circumstances. The emphasis is on empirical evidence, studying rule-based action and recognising the historical, social and interpretative contexts. Routines are defined in this project as established, significant, sanctioned and recurrent practices within organisations, and operational and strategic routines can be identified in each of the companies. Changes in routines can be explained using a recursive form of analysis.

The discussion has emphasised the relationship between structure and agency, drawing on the firms' culture and producing particular routines.

Routines are seen as practices built on the meanings defined in the culture and the distinction between practice and meaning has been valuable.

What have we found? Three of the companies have disappeared: two, Unipalm-Pipex and Chadwyck-Healey, through mergers or take-overs which demonstrate their success. They were eagerly sought by other firms wishing to acquire their expertise. The third, Digital, was acquired by Compaq as a 'best of breed partnership' according to the Compaq press release (11 June 1998) which accompanied it, but was a take-over of an ailing giant, with a highly regarded product portfolio, rather than the acquisition of lively new directions. At the time of writing, the National Extension College retains its niche and continues to prosper. The single most notable feature of the three companies which have thus demonstrated greater elements of success is the extent to which they were alert to and sought out new opportunities. And this feature of their behaviour was, in each case, embedded in the taken-for-granted ways in which they worked. It was not an additional factor to be considered after other matters had been decided or solely the province of a separate department. Each of the companies had a sense of what it was doing but not precisely of where it was going. But their alertness to the environment they inhabited enabled them to adjust as circumstances changed.

## Objectives revisited

### A *richer social ontology*

Routines are difficult to study because they are complex patterns of social action. They have been characterised in the literature both as automatic responses and as accomplishments which have to be worked at continually. The evidence here demonstrates that they are quite clearly more than automatic responses and that broadening the use of the term to include accomplishments acquired by effort throughout organisations helps to explain organisational behaviour.

The implication of denying this claim and retaining a narrow definition of routine is that any behaviour which is not automatic, or not very largely automatic, must be either relatively carefully thought through or randomly responsive. The evidence of the four cases shows that we can reject that construction. Throughout organisations we find behaviour which is taken for granted but not automatic. Low-level and high-level work are both prone to such practices. The behaviour is routine in the sense that it is sanctioned, significant and established, and it is recurrent in a path-dependent way. The regular practices of the organisations studied were often responsive or intuitive in manner, in ways which many analysts of strategic change have outlined (Mintzberg 1978; Johnson 1987), but they also had a regularity and predictability about them arising from the situated nature of the events under study. There were particular histories, social, political and organisational patterns from which the events arose and through which they were interpreted. In order to comprehend fully the behaviour of firms it is important therefore to recognise the routineness of practices throughout the firm.

Using the concept of routine, as defined above, enriches our social ontology. It emphasises the importance of an institutional understanding of firms' behaviour and gives us the potential for a rich picture throwing new light on operational and strategic behaviour and the process of change.

### Routines and change

Routineness in the organisations studied has not been inimical to change. Some routine practices may be constraining in novel circumstances. The outcomes are open, however: they may or may not encourage change and the change may or may not be helpful. The processes discussed in each case evolved from earlier practices. Such evolution is not necessarily benign. It has the potential to damage the firm's future or to enhance it but it is clear that firms and organisations can develop routines which positively encourage change.

### Routines and technology

Technology was implicated in the changes analysed. The technology under consideration was electronic and was associated with the provision or analysis of information. Technology, in the cases discussed here, was (partly) *constitutive* of the firms; that is, it had the power to enact or establish the firms. It did not do this on its own but, in interacting with other factors, it had major implications for the structure and routine behaviour of the firms. This argument suggests that other factors on their own, working without technology, would result in a firm which was significantly different from the case observed. The view of information technology as a tool which is relevant for the behaviour of the firm only to the extent to which human agency gives it power is inadequate on these arguments. Information systems are directly implicated in the flow of information, the opportunities available for acquiring knowledge, the codification of that knowledge and its reproduction and storage. Information technology affects shared meanings since it is directly involved in passing on some of those meanings in its day-to-day work. The cases set out in the previous chapters suggest that technology has implications for human agency which go beyond the initial intentions of the agents. Technology is not determining, but interacts with the other factors identified in a recursive way; it cannot be adequately analysed, therefore, outside the social system of which it is a part.

## Explanations and evidence

The approach adopted here has argued:

- that the persistence of social structures is dependent on practices, some of which are routine, which the structures help to constitute;
- that social structures are only *relatively* enduring;
- that structures and practices have an existence which is largely independent of any one individual;
- that the practices, in one way or another, are governed, conditioned, limited and facilitated by the structures but cannot be reduced solely to them. Action cannot be reduced to structure and structure cannot be reduced to action.

In looking for causal factors and for the explanation of why some routines endure and some evolve, we must recognise that the relevant factors may be unobservable. There are tendencies and structures lying behind events and their patterns. The explanation of routines here has adopted the position that they endure to some extent independently of our knowledge of them. Such an approach has been helpful in addressing, in particular, the material nature of technology.

Thus, in the companies analysed, the appropriate interpretative framework has been neither technologically determinist nor socially determinist. The material and social features of the cases have interacted. Human action has been constrained and enabled by the technology, and the technology has been interpreted and changed by human agency.

The project sought out patterns in behaviour, recognising that such patterns may be influenced by a range of factors and events and that actual outcomes are open even though there may be a regularity in the way in which the firms studied behaved. Such an approach called for a research method which was able to investigate deep structures and the practices associated with them. The use of grounded theory to frame the research method enabled that kind of investigation to take place.

This required a continual, relentless rewriting of the stories of each company and was central to the development of explanatory patterns. In order to avoid the risk that theoretical explanations became little more than figments of the imagination, the stories had to be regularly checked back against the data. The explanatory framework was continually regrounded. Critical realist categories slowly emerged as a helpful orientation though it was clear that the distinctions in many critical realist accounts were sharper or more subtle than could be made by observation.

A critical realist frame of reference and a grounded research method are complementary. Each is sympathetic to the precepts of the other.

Lawson (1997) sets out the situated rationality through which, he argues, we can better understand human agency:

Not only are individuals' choices of actions conditioned by the situated options which they perceive, but also the individuals themselves, their expressions of their needs and motives, the manner in which their capacities and capabilities have been moulded, their values and interests and so forth, are conditioned by the context of their birth and development. At any point in time any individual is situated in a range of positions, with associated, perhaps contradictory, real interests, as well as other needs and motives. Associated with the positions in which any individual is located will be a range of rules to draw upon, obligations to fulfil, structures of power to utilise and be influenced by. Many such social structures will be inadequately or falsely understood. Most of the skills utilised, modes of conduct performed, will be tacit. Action in such a context is a continuous stream, continuously monitored, and rarely rendered available to discourse.

(Lawson 1997, p. 187)

The evidence we have from the cases discussed here is that the individual respondents and the institutions for which they worked were heavily influenced by structures and rules situated in the historical and socio-economic context in which the firms existed, 'their birth and development'. The rules were operated by individuals but in a context which enabled us to perceive particular institutional (Digital, NEC, Unipalm-Pipex or Chadwyck-Healey) ways of doing things. Through analysis of the interpretations of institutional members and the observation and scrutiny of institutional artefacts, including information technologies, (some of) the action *is* 'rendered available to discourse', giving us enriched conceptual categories through which the better to understand the behaviour of firms.

# Bibliography

Alchian, A.A. (1950) 'Uncertainty, evolution and economic theory', *Journal of Political Economy*, LVIII, 3: 211–21.

Altheide, D.L. and Johnson, J.M. (1994) 'Criteria for assessing interpretive validity in qualitative research', in N.K. Denzin and Y.S. Lincoln (eds) *Handbook of Qualitative Research*, London: Sage.

Andersen, E.S. (1994) *Evolutionary Economics: Post Schumpeterian Contributions*, London: Pinter Publishers.

Argyris, C. (1976) *Increasing Leadership Effectiveness*, New York: John Wiley & Sons.

Baert, P. (1996) 'Realist philosophy of the social sciences and economics: a critique', *Cambridge Journal of Economics*, 20, 5: 513–22.

Barrie, C., Beavis, S. and Tran, M. (1997) 'WorldCom knocks BT off global expansion line', *Finance Guardian*, 11 November, p. 19.

Becker, M. (1998) 'An empirical contribution to a taxonomy of routines', unpublished MPhil dissertation, Judge Institute of Management Studies, University of Cambridge.

Bhaskar, R. (1979) *The Possibility of Naturalism*, Hemel Hempstead: Harvester Press.

Blackie, J. (1970) *Report of an Inspection of the National Extension College, Cambridge*, Cambridge: National Extension College.

Boland, R.J. (1985) 'Phenomenology: a preferred approach to research on information systems', in E. Mumford, R. Hirschheim, G. Fitzgerald and T. Wood-Harper (eds) *Research Methods in Information Systems*, New York: North-Holland.

—— (1996) 'Why shared meanings have no place in structuration theory: a reply to Scapens and Macintosh', *Accounting, Organization and Society*, 21, 7/8: 691–7.

Boulding, K.E. (1981) 'The basic evolutionary model', *Evolutionary Economics*, Beverly Hills, Calif., and London: Sage; reprinted in U. Witt (1993) *Evolutionary Economics*, The International Library of Critical Writings in Economics 25, Aldershot: Edward Elgar.

Bryman, A. (1989) *Research Methods and Organization Studies*, London: Unwin Hyman.

Chadwyck-Healey (1996) Press release, November, on-line, available HTTP:http://www.chadwyck.co.uk (December 1996).

Chandler, A.D., Jr (1962) *Strategy and Structure: Chapters in the History of Industrial Enterprise*, Cambridge, Mass.: MIT Press.

—— (1977) *The Visible Hand: the Managerial Revolution in American Business*, Cambridge, Mass.: The Belknap Press of Harvard University Press.

—— (1988) *Scale and Scope*, Cambridge, Mass.: Harvard University Press.

Child, J. (1972) 'Organizational structure, environment and performance: the role of strategic choice', *Sociology*, 6, 1: 1–22.

Child, J. and Smith, C. (1987) 'The context and process of organizational transformation', *Journal of Management Studies*, 24, 6: 565–94.

Cohen, M.D. and Bacadayan, P. (1994) 'Organizational routines are stored as procedural memory: evidence from a laboratory study', *Organization Science*, 5, 4 November: 554–68.

Compaq (1998) 'Digital shareholders approve acquisition of Compaq', press release, 11 June, on-line, available HTTP: http://www.compaq.com/newsroom/pr/1998/pr110698a.html (28 January 2000)

Cronin, M.J. (1994) *Doing Business on the Internet*, New York: Van Nostrand Reinhold.

David, P.A. (1985) 'Clio and the economics of QWERTY', *American Economic Review*, 75, 2: 332–7; reprinted in U. Witt (1993) *Evolutionary Economics*, The International Library of Critical Writings in Economics 25, Aldershot: Edward Elgar.

de Salvo, A. (1993) *Kett of Cambridge; an Eminent Victorian and his Family*, Cambridge: National Extension College.

Digital (1992) *Flexible Work Practices: an Overview*, Basingstoke: Digital Equipment Company Ltd.

—— (1993) *Digital Equipment Corporation*, 1993 Annual Report, Maynard, Mass.: Digital Equipment Corporation.

—— (1994) 'Flexible working handbook', Central PSC, Digital Equipment Company Ltd, Newmarket (unpublished internal document).

—— (1995) *Digital Brief Guide*, Maynard, Mass.: Digital Equipment Corporation.

Douglas, M. (1987) *How Institutions Think*, London: Routledge and Kegan Paul.

Du Gay, P., S. Hall, L. Janes and K. Negus (eds) (1997) *Doing Cultural Studies: the Story of the Sony Walkman*, London: Sage.

Dutton, J.E. (1988) 'Understanding strategic agenda building and its implications for managing change', in L.R. Pondy, R.J. Boland Jr, and H. Thomas (eds) *Managing Ambiguity and Change*, Chichester: Wiley.

Eisenhardt, K.M. (1989) 'Building theories from case study research', *Academy of Management Review*, 14, 4: 532–50.

Freeman, R. (1983) *The National Extension College, Cambridge: its Aims and Development*, Milton Keynes: Open University.

Garnsey, E. (1992) 'Towards a critical systems perspective; on constitutive processes in dynamic social systems', unpublished paper, Management Studies Group, Department of Engineering, University of Cambridge.

Giddens, A. (1979) *Central Problems in Social Theory: Action, Structure and Contradiction in Social Analysis*, London: Macmillan.

—— (1982) *Profiles and Critiques in Social Theory*, London: Macmillan.

—— (1984) *The Constitution of Society*, Cambridge: Polity.

—— (1990) 'Structuration theory and sociological analysis', in J. Clark, C. Modgil and S. Modgil (eds) (1990) *Anthony Giddens: Consensus and Controversy*, London: Falmer.

Ginsberg, M.J. (1979) 'A study of the implementation process', in R. Doktor, R.L.

Schultz and D.P. Slevin (eds) *TIMS Studies in the Management Sciences*, 13, Amsterdam: North Holland.

Glaser, B.G. and Strauss, A.L. (1967) *The Discovery of Grounded Theory: Strategies for Qualitative Research*, Chicago: Aldine.

Goddard, A. (1999) 'Knight's errand', *The Times Higher*, 15 October.

Grant, R.M. (1991) 'The resource-based theory of competitive advantage: implications for strategy formulation', *California Management Review*, Spring: 114–35.

Grinyer, P.H. and Spender, J.C. (1979) 'Recipes, crises and adaptation in mature businesses', *International Studies of Management and Organizations*, IX, 3: 113–33.

*Guardian* (1996) 'UK net loses pioneer', *Guardian Online*, 23 May.

Hall, S. (1997) 'The work of representation', in S. Hall (ed.) *Representation: Cultural Representations and Signifying Practices*, London: Sage.

Hari Das, T. (1983) 'Qualitative research in organizational behaviour', *Journal of Management Studies*, 20, 3: 301–14.

Hirschman, A.O. (1970) *Exit, Voice and Loyalty*, Cambridge, Mass.: Harvard University Press.

Hodgson, G.M. (1997) 'The ubiquity of habits and rules', *Cambridge Journal of Economics* 21, 6: 663–84.

Johnson, G. (1987) *Strategic Change and the Management Process*, Oxford: Blackwell.

—— (1988) 'Rethinking incrementalism', *Strategic Management Journal*, 9: 75–91.

Jones, M.R. (1994) 'Organizational learning: collective mind or cognitivist metaphor?', unpublished paper, Management Studies Group, Department of Engineering, University of Cambridge.

—— (1998a) 'Structuration theory', in W.J. Currie and R. Galliers (eds) *Rethinking Management Information Systems*, Oxford: Oxford University Press.

—— (1998b) 'Steering between the Scylla of embedded structure and the charybdis of strong symmetry: an information systems Odyssey', unpublished paper, Judge Institute of Management Studies, University of Cambridge.

Joshi, K. (1991) 'A model of users' perspectives on change: the case of information systems technology implementation', *MIS Quarterly*, 15, 2: 229–42.

Keat, R. and Urry, J. (1975) *Social Theory as Science*, London: Routledge and Kegan Paul.

Keen, P.G.W. (1981) 'Informational systems and organizational change', *Communications of the ACM*, 24, 1: 24–33.

Kling, R. and Iacono, S. (1984) 'The control of information systems development after implementation', *Communications of the ACM*, 27, 12: 1218–26.

Latour, B. (1996a) *Aramis or the Love of Technology*, Cambridge, Mass.: Harvard University Press.

—— (1996b) 'Social theory and the study of computerized work sites'. in W.J. Orlikowski, G. Walsham, M.R. Jones and J.I. DeGross (eds) *Information Technology and Changes in Organizational Work*, London: Chapman and Hall.

Lawson, T. (1997) *Economics and Reality*, London: Routledge.

—— (1999) 'Developments in economics as realist social theory' in S. Fleetwood (ed.) *Critical Realism in Economics; Development and Debate*, London: Routledge.

Layder, D. (1987) 'Key issues in structuration theory: some critical remarks', *Current Perspectives in Social Theory*, 8: 25–46.

Lewin, K. (1952) 'Group decision and social change', in T. Newcombe and E. Hartley (eds) *Readings in Social Psychology*, New York: Holt.

Lewis, D. (1968) *Convention: a Philosophical Study*, Cambridge, Mass.: Harvard University Press.

Lewis, P. (1999) 'Metaphor and critical realism', in S. Fleetwood (ed.) *Critical Realism in Economics; Development and Debate*, London: Routledge.

Leydesdorff, L. and Van Den Besselaar, P. (eds) (1994) *Evolutionary Economics and Chaos Theory; New Directions in Technology Studies*, London: Pinter.

Loasby, B.J. (1991) *Equilibrium and Evolution: an Exploration of Connecting Principles in Economics*, Manchester: Manchester University Press.

Lundvall, B.A. (ed.) (1992) *National Systems of Innovation: Towards a Theory of Innovation and Interactive Learning*, London: Pinter.

Lyttinen, K. and Hirschheim, R. (1987) 'Information systems failures: a survey and classification of the empirical literature', *Oxford Surveys in Information Technology*, 4: 257–309.

McCraw, T.K. (ed.) (1988) *The Essential Alfred Chandler: Essays Toward a Historical Theory of Big Business*, Boston, Mass.: Harvard Business School Press.

March, J.G. and Simon, H.A. (1958) *Organizations*, New York: Wiley.

March, J.G. and Simon, H.A. (1993) 'Organizations revisited', *Industrial and Corporate Change*, 2, 3: 299–316.

Markus, M.L. (1983) 'Power, politics and MIS implementation', *Communications of the ACM*, 26, 6: 430–44.

Matthews, R.C.O. (1984) 'Darwinism and economic change', in D.A. Collard, D.R. Helm, M.F.G. Scott and A.K. Sen (eds) *Economic Theory and Hicksian Themes*, Oxford: Clarendon Press, reprinted in U. Witt (1993) *Evolutionary Economics*, The International Library of Critical Writings In Economics 25, Aldershot: Edward Elgar.

Miller, D. (1982) 'Evolution and revolution: a quantum view of structural change in organizations', *Journal of Management Studies*, 19, 2: 131–51.

Miller, D. and Friesen, P.H. (1980) 'Momentum and revolution in organizational adaptation', *Academy of Management Journal*, 23, 4: 591–614.

Mintzberg, H. (1978) 'Patterns in strategy formation', *Management Science*, 24, 9: 934–48.

—— (1990) 'The design school: reconsidering the basic premises of strategic management', *Strategic Management Journal*, 11, 3: 171–95.

Mintzberg, H. and Quinn, J.B. (1991) *The Strategy Process: Concepts, Contexts, Cases*, 2nd edn, Englewood Cliffs, NJ: Prentice-Hall International.

Mintzberg, H. and Waters, J.A. (1982) 'Tracking strategy in an entrepreneurial firm', *Academy of Management Journal*, 25, 3: 465–99.

Mirowski, P. (1987) 'The philosophical bases of institutional economics', *Journal of Economic Issues*, 21, 3: 1001–38.

Morgan, G. (1986) *Images of Organization*, Beverly Hills, Calif.: Sage.

—— (1989) 'Do organizations enact their environments?', in G. Morgan, *Creative Organization Theory*, Beverly Hills, Calif.: Sage.

Nadler, D.A. and Tushman, M.L. (1990) 'Beyond the charismatic leader: leadership and organizational change', *California Management Review*, Winter: 77–97.

National Extension College (1990) *The National Extension College: a Catalyst for Educational Change*, Cambridge: National Extension College.

—— (1996) Reports of internal reviews: R. Webb, 'Organisational review project'; J. Bell and S. Nicholls, 'Quality framework'; G. Belfiore, 'Changing environments for learning'; internal unpublished papers, Cambridge: National Extension College.

Nelson, R.R. (1995) 'Recent evolutionary theorising about economic change', *Journal of Economic Literature*, 33: 48–90.

Nelson, R.R. and Winter, S.G. (1982) *An Evolutionary Theory of Economic Change*, Cambridge, Mass.: The Belknap Press of Harvard University Press.

Ngwenyama, O.K. (1998) 'Groupware, social action and organizational emergence: on the process dynamics of computer mediated distributed work', unpublished paper, Virginia Commonwealth University, USA, and Aalborg University, Denmark.

North, D.C. (1990) *Institutions, Institutional Change and Economic Performance*, Cambridge: Cambridge University Press.

Orlikowski, W.J. (1992) 'The duality of technology: rethinking the concept of technology in organizations', *Organization Science*, 3, 3: 398–427.

—— (1995) 'Action and artifact: the structuring of technologies-in-use', unpublished paper, Sloan School of Management, Cambridge, Mass.: Massachusetts Institute of Technology.

Penrose, E.T. (1952) 'Biological analogies in the theory of the firm', *American Economic Review*, 62, 5: 804–19.

—— (1959) *The Theory of the Growth of the Firm*, Oxford: Basil Blackwell.

Pentland, B.T. and Rueter, H.H. (1994) 'Organizational routines as grammars of action', *Administrative Science Quarterly*, 39, 3: 484–510.

Pettigrew, A. (1985) *The Awakening Giant*, Oxford: Blackwell.

Pfeffer, J. (1981) 'Management as symbolic action', *Research in Organizational Behaviour*, 3: 1–52.

Pickering, A. (1995) *The Mangle of Practice: Time, Agency and Science*, Chicago: University of Chicago Press.

Quinn, J.B. (1980) *Strategies for Change: Logical Incrementalism*, Holmwood, Ill.: Irwin.

Samuels, W.J. (1995) 'The present state of institutional economics', *Cambridge Journal of Economics*, 19, 4: 569–90.

Schofield, J. (1996) 'Survival of the frittest', *Guardian*, 21 November.

Schultz, R.L. and Slevin, D.P. (1979) 'Introduction: the implementation problem', in R. Doktor, R.L. Schultz and D.P. Slevin (eds) *TIMS Studies in the Management Sciences*, 13, Amsterdam: North-Holland.

Senge, P.M. (1990) *The Fifth Discipline: the Art and Practice of the Learning Organization*, New York: Doubleday.

Spender, J.C. (1989) *Industry Recipes*, Oxford: Blackwell.

Srinivasan, A. and Davis, J.G. (1987) 'A re-assessment of implementation process models', *Interfaces*, 17, 3: 64–71.

Stinchcombe, A. (1990) 'Milieu and structure updated: a critique of the theory of structuration' in J. Clark, C. Modgil and S. Modgil (eds) *Anthony Giddens: Consensus and Controversy*, London: Falmer.

Strassmann, D, and Polanyi, L. (1995) 'The economist as storyteller', in E. Kuiper and J. Sap (eds) *Out of the Margin: Feminist Perspectives on Economics*, London: Routledge.

Strauss, A.L. (1987) *Qualitative Analysis for Social Scientists*, Cambridge: Cambridge University Press.

Symons, V. (1990) 'Evaluation of information systems: multiple perspectives', unpublished PhD thesis, University of Cambridge.

Teece, D.J., Pisano, G. and Shuen, A. (1990) 'Firm capabilities and the concept of strategy', CCC Working Paper No. 90–8, University of California at Berkeley: Consortium on Competitiveness and Co-operation, Center for Research in Management.

Tesch, R. (1990) *Qualitative Research: Analysis Types and Software Tools*, Basingstoke: Falmer.

Thomas, B. (1991) 'Alfred Marshall on economic biology', *Review of Political Economy*, 3, 1: 1–14.

Tushman, M.L. and Romanelli, E. (1985) 'Organizational evolution: a metamorphosis model of convergence and reorientation', in L.L. Cummings and B.M. Staw (eds) *Research in Organizational Behavior*, 7, London: JAI Press.

Unipalm Group (1995) *Annual Report and Accounts*, Cambridge: Unipalm Group.

Van Maanen, J. (1979a) 'Reclaiming qualitative methods for organizational research: a preface', *Administrative Science Quarterly*, 24, December: 520–6.

—— (1979b) 'The fact of fiction in organizational ethnography', *Administrative Science Quarterly*, 24, December: 539–49.

—— (1988) *Tales of the Field: on Writing Ethnography*, London: University of Chicago Press.

Walsham, G. (1991a) 'Organizational metaphors and information systems research', *European Journal of Information Systems*, 1, 2: 83–94.

—— (1991b) 'Implementation issues in information systems assessment', published as 'Problèmes d'implémentation dans l'évaluation des systèmes d'information', *Technologies de l'Information et Société*, 3, 2: 69–85.

—— (1993) *Interpreting Information Systems in Organizations*, Chichester: John Wiley and Sons.

—— (1997) 'Actor-network theory and IS research: current status and future prospects', in A.S. Lee, J. Liebenau and J.I. DeGross (eds) (1997) *Information Systems and Qualitative Research*, London: Chapman and Hall.

Walsham, G. and Han, C.K. (1993) 'Information systems strategy formation and implementation: the case of a central government agency', *Accounting, Management and Information Technology*, 3, 3: 191–209.

Weick, K.E. (1979) *The Social Psychology of Organising*, 2nd edn, Reading, Mass.: Addison Wesley.

Williamson, O.E. (1975) *Markets and Hierarchies: Analysis and Antitrust Implications*, New York: The Free Press.

Woolf, V. (1926) *Diaries*, vol. 3, p. 74, quoted in H. Lee (1996) *Virginia Woolf*, London: Chatto and Windus.

Zuboff, S. (1988) *In the Age of the Smart Machine*, Oxford: Heinemann.

# Index